Literary Criticism and Cultural Theory

Edited by

William E. Cain

Professor of English

Wellesley College

A ROUTLEDGE SERIES

LITERARY CRITICISM AND CULTURAL THEORY

WILLIAM E. CAIN, *General Editor*

MUSEUM MEDIATIONS

Reframing Ekphrasis in
Contemporary American Poetry

Barbara K. Fischer

Routledge
New York & London

Published in 2006 by
Routledge
Taylor & Francis Group
711 Third Avenue,
New York, NY 10017

Published in Great Britain by
Routledge
Taylor & Francis Group
2 Park Square,
Milton Park, Abingdon,
Oxfordshire OX14 4RN

First issued in paperback 2014

Routledge is an imprint of the Taylor and Francis Group, an informa business

International Standard Book Number-13: 978-0-415-97534-6 (Hardcover)
Library of Congress Card Number 2005031224

Library of Congress Cataloging-in-Publication Data

Fischer, Barbara K., 1971-
　　Museum mediations : reframing ekphrasis in contemporary American poetry / by Barbara K. Fischer.
　　　　p. cm. -- (Literary criticism and cultural theory)
　　Includes bibliographical references and index.
　　ISBN 978-0-415-97534-6 (hbk)
　　ISBN 978-1-138-01172-4 (pbk)
　　1. American poetry--20th century--History and criticism. 2. Museums in literature. 3. Ekphrasis.
I. Title. II. Series.

PS310.M76F57　2005
811'.5093--dc22
　　　　　　　　　　　　　　　　　　　　　　　　　　　　　　　　　　　　　2005031224

Taylor & Francis Group
is the Academic Division of Informa plc.

Visit the Taylor & Francis Web site at
http://www.taylorandfrancis.com

and the Routledge Web site at
http://www.routledge-ny.com

for James and Emily

Contents

List of Figures

Acknowledgments

I am grateful to Denis Donoghue for his insight, encouragement, and attentive commentary as I wrote this book. Phillip Brian Harper offered crucial feedback at every stage in the study's evolution, and Bruce Altshuler generously took on the role of guide to museum studies. Their enthusiasm and expertise have been greatly appreciated. This study has also benefited from conversations about poetry, art, and ekphrasis with Richard Howard, Elizabeth Bergmann Loizeaux, Ernest Gilman, and Mary Jo Bang. A discussion with W. J. T. Mitchell when I was beginning my research spurred my efforts to find new ways to value ekphrastic poetry. I am especially grateful to the late Kenneth Koch for timely correspondence that pointed me in the direction of several key examples. My ongoing conversation with Timothy Donnelly is a source of catalyzing ideas and much pleasure. Sarah Stevenson, Karine Schaefer, Loretta Mijares, and Anne Green, colleagues at NYU, provided a needed forum for ideas at several turning points, and Max Novick at Routledge helped see this into print. Special thanks to Troy Thibodeaux for his countless insights and tireless editing over many years, and most of all for his camaraderie in this pursuit.

While writing this book, I have had the abundant good fortune to have a supportive community of family and friends to see me through with good sense and companionship on the homefront. This book would not have been possible without the friendship and hands-on help of Erin Donovan, Michelle Stratman, Liz Rennick, Marcie Hille, Jennifer Green, Kinga Toader, Michele Patterson, and the TTNS and Rivertown moms. Nor would it have been possible without the immeasurable encouragement and assistance of my family—it is to my siblings and siblings-in-law, to Jane and Dennis Allendorf, to Joseph Cassidy, and most especially to my parents, Peter and Donna Fischer, that I owe the most. My deepest thanks are due to John Allendorf for his patience, affection, and wit, and for his pilgrim soul. This book is dedicated to our children Jamie and Emily, whose laughter has accompanied its writing.

COPYRIGHT PERMISSIONS

From "Nydia, The Blind Flower Girl of Pompeii," by Laurence Goldstein, in *A Visit to the Gallery*, ed. Richard Tillinghast (The University of Michigan Museum of Art, 1997). Used by permission.

Barbara Guest, "The Poetess," from *Moscow Mansions*, published by Viking/Doubleday in 1973, © 1973 by Barbara Guest. Forthcoming in *Collected Poems of Barbara Guest*, from Wesleyan University Press. Used by permission of Wesleyan University Press and the author.

"Monet's 'Waterlilies.'" Copyright © 1970 by Robert Hayden, from *Collected Poems of Robert Hayden* by Robert Hayden, edited by Frederick Glaysher. Used by permission of Liveright Publishing Corporation.

Reprinted by permission of Farrar, Straus and Giroux, LLC: Excerpts from *Inner Voices: Selected Poems, 1963–2003*. Copyright © 2004 by Richard Howard.

Excerpts from poems by Richard Howard that are not collected in *Inner Voices*, from *The Damages*, *Trappings*, and *Talking Cures*, are used by permission of the author. Copyright © 1967, 1999, 2002, by Richard Howard.

"Dialogue," from *At the Palace of Jove* by Karl Kirchwey. Copyright © 2002 by Karl Kirchwey. Used by permission of Marion Wood Books, an imprint of G. P. Putnam, a division of Penguin Group (USA) Inc.

Excerpts from the works of Kenneth Koch used by permission of the Kenneth Koch Literary Estate.

"The Sea, That Has No Ending . . . ," from *Passing Through: The Later Poems New and Selected* by Stanley Kunitz. Copyright © 1995 by Stanley Kunitz. Used by permission of W. W. Norton & Company, Inc.

Ken Mikolowski, "You Are What You Art," from *A Visit to the Gallery*, University of Michigan Museum of Art. Used by permission.

Excerpt from "A Visit to the Gallery" by Molly Peacock quoted with permission of the author.

The lines from "Love in the Museum." Copyright © 1993, 1955 by Adrienne Rich, from *Collected Early Poems: 1950–1970* by Adrienne Rich. Used by permission of the author and W. W. Norton & Company, Inc.

From *The Collected Poems of Wallace Stevens* by Wallace Stevens, copyright © 1954 by Wallace Stevens and renewed 1982 by Holly Stevens. Used in North America by permission of Alfred A. Knopf, a division of Random House, Inc. and in the UK by permission of Faber and Faber Ltd.

Excerpts from Cole Swensen, *Try*, and *Such Rich Hour*, published by the University of Iowa Press. Used by permission.

Diane Wakoski, "Old Embroidered Chinese Robes in the Ann Arbor Museum." Copyright © Diane Wakoski. Used by permission.

Excerpts from "Spring (The Procession)," by Rachel Wetzsteon, reprinted by permission of the author.

Excerpts from Susan Wheeler, *Ledger*, published by the University of Iowa Press. Used by permission.

"Poem for museum goers," by John Wieners, from *John Wieners: Selected Poems 1958–1984*, edited by Raymond Foye (Santa Barbara, CA: Black Sparrow Press, 1986). Copyright © 1986 by John Wieners. Reprinted by permission of Black Sparrow Books, an imprint of David R. Godine, Publisher, Boston.

Reprinted by permission of Farrar, Straus and Giroux, LLC: Excerpts from "Summer Storm" from *Negative Blue: Selected Later Poems* by Charles Wright. Copyright © 2000 by Charles Wright.

Grateful acknowledgment to the following for permission to reprint the following images:

Jules Breton, *The Song of the Lark,* 1884. Photography copyright © The Art Institute of Chicago.

René Magritte, *The Old Gunner,* 1947. Copyright © 2005 C. Herscovici, Brussels / Artists Rights Society (ARS), New York.

Introduction

The Museum as Muse

In 1999, the Museum of Modern Art presented the exhibition *The Museum as Muse: Artists Reflect,* assembling the works of 60 artists who took the museum as their subject. From Roger Fenton's 1857 photograph of copyists in the British Museum's gallery of antiquities, to Claes Oldenburg's *Mouse Museum* (1965–77), to a video of Andrea Fraser's performance as mock-docent at the Wadsworth Atheneum (1991), the show displayed artists' engagements with an institution central to their education, career development, and canonization. Their attitudes toward it ranged from willing collaboration to wary cohabitation to outright subversion: as Kynaston McShine writes in the catalog introduction, the contents of museums inspire curiosity and wonder, "But artists of this century have shown a desire to explore the frame within which that sense of wonder is maintained" (17). Taking the museum as topic gave artists opportunities for critique—of the museum's collecting practices and patronage structures, for example—though the exhibit testified to the ease with which those critiques could be absorbed, arranged in "a general narrative sequence" (25) on MoMA's walls. Yet even in a neutralizing survey, these glimpses of the museum in the museum allowed for startling recognition of the conventions shaping experiences of art, conventions we normally overlook. Thomas Struth's 1994 photograph of MoMA visitors catches them in blurred motion in front of Jackson Pollock's *One: Number 31, 1950*—leaning in with a squint, turning away in bewilderment, clutching an explanatory brochure, taking a snapshot as souvenir. Struth explains that he strives to set up a disquieting encounter for the museumgoer who contemplates his photograph of museumgoers: "Therein lies a moment of pause or of questioning" (qtd. in McShine 17).

This study turns to such moments of pause and questioning in the work of contemporary poets. It assembles an analogous collection of responses to the museum in poetry, analyzing poems that foreground and interrogate an art museum setting. Poets, like artists, approach the museum as an arena of perception, but because the museum is not their primary institution of canonization (though they benefit from museum sponsorship, as I discuss in chapter one), they stand at a further remove. They enter the provinces of the visual arts as observers, admirers, and interlopers, and they often tell about their encounters through an established mode in their own medium—ekphrasis. This term for the traditional subgenre of poems about works of art derives from a rhetorical trope for vivid description, for "telling in full" (from *ekphassein*, to speak forth): poets describe or envoice works of visual art by summoning the pliability and eloquence of their own medium.[1] As James Heffernan defines the mode, ekphrasis is "the verbal representation of visual representation" (3), an exchange by which one medium is translated into the signs of another. I would expand this definition to reflect the fact that contemporary ekphrastic[2] poems may address non-representational visual works, or may not "represent" their subjects at all, riffing off their visual sources more tangentially or interrogatively. Ekphrasis in contemporary poetry is, to employ Cole Swensen's practical definition, "the product of a writer's contemplation before a painting, sculpture, Grecian urn, Achillean shield, or other specific work of art" ("To Writewithize" 122).[3] That product may take a variety of forms, from a sonnet to a collage of verbal fragments, and that act of contemplation may be sparked by images in books or magazines, on billboards, stages, or screens, but for contemporary poets standing "before" a work of art most often situates them in a museum.

In that location, from that perspective of an outsider's distance in the precincts of the visual arts, ekphrasis is an interpretive occasion and a critical tool—a mode that involves description, enumeration, analysis, comparison, citation, questioning, critique, assessment, summation, and judgment. My emphasis on ekphrasis as a form of critical work departs from the two dominant approaches to the mode. In one, ekphrasis is associated with the pictorialist tradition in poetry, the tradition of *ut pictura poesis* ("as in painting, so in poetry") in which the "sister arts" are reciprocally inspiring.[4] Theories of this affiliation, such as Murray Krieger's, are concerned with ekphrasis as a dialectical movement between images and texts—a narrativizing and temporalizing of visual stasis, and a stilling of verbal-temporal movement—that is synthesized in the "still movement" of the verbal icon. A second approach, which has its roots in da Vinci's *Paragone*, or competition between the arts, and in G. E. Lessing's insistence on the separate domains of poetry and painting, emphasizes that images and texts are pitted against each other in representational and ideological conflicts. W. J. T. Mitchell's studies of the political underpinnings of a culture's relative iconophobia and iconophilia draw on this understanding of images and texts in paragonal struggle. Rather than focusing on ekphrasis as either fertile reciprocity or fight for dominance (though both moods do color the poems I discuss), I present ekphrasis as a form of critical mediation. The poet in the museum approaches the visual arts from an angle of displacement that invites a mix of commentary, homage, resistance, argument, and self-criticism. If poets press the advantage of their medium, it is to develop language's discursive and critical capacities in ways that nonetheless concede the priority of the visual work. From this position—*after* art but as its spokesperson—poets use ekphrasis to allow for an interplay of complicity and provocation.

In the process of doing this critical work, some poets catch the museum in the corner of the eye—a security guard in a doorway, a wall text or ticket booth, an effect of the lighting. This study presents ekphrastic poems that exhibit "peripheral vision," an acute awareness of the physical and institutional conditions that frame encounters with art. The museum setting has been increasingly important to ekphrasis since the Romantic period, and twentieth-century poetry reflects the institution's pervasive influence. As Elizabeth Bergmann Loizeaux observes, "with the growth over the past 200 years of the museum as the primary place for viewing works of art, the experience of museum culture with all of its institutional apparatus, including reproductions and art-historical commentary, now typically informs the writing of ekphrastic poetry" (80). Heffernan traces this influence from Keats's sonnet on the Elgin marbles (1817) to Auden's "Musée des Beaux Arts" (1938) to Ashbery's "Self-Portrait in a Convex Mirror" (1974), arguing that ekphrasis as a self-contained lyric mode emerged as a result of the museum's presentation of individual artworks for scrutiny (135–89). Catherine Paul has described the particular relevance of museums and exhibitionary methods to high modernist poetry.[5] My analysis departs from these critics' work chronologically (I begin at midcentury), and also in that I am concerned primarily with the ways poets themselves notice and scrutinize the institutional apparatus of the museum—the museum appears in the poems directly as subject matter. We are reminded of the museum in titles throughout the century—Yeats's "The Municipal Gallery Revisited," Frank O'Hara's "On Seeing Larry Rivers' *Washington Crossing the Delaware* at the Museum of Modern Art," Anthony Hecht's "At the Frick," Rita Dove's *Museum*—but these poets, like most of their ekphrastic forebears, quickly darken their peripheral vision to focus on concerns and themes intrinsic to the works at hand. This study addresses those who sustain their sidelong glances.

In a series of close readings, I argue that the peripheral vision we find in later twentieth-century ekphrastic poetry, the attentiveness to the museum setting, reflects poets' need to find a critical, revisionary stance that does not disavow the force of aesthetic perception. These poets indict the museum on several counts: for the sacralizing aura that obfuscates the conditions of a work's making, for elitism and escapism, and for the insularity that divorces art from everyday life and the political sphere. But they also turn to museums for aesthetic pleasure, sensory intensity, evocative detail, visual and verbal harmonies and contrasts, figures for desire—for the range of emotional and intellectual attractions of art apprehended as such. The museum captures poets in an ambivalent position between the pleasures of aesthetic experience and the skepticism of institutional critique. For this reason, I will show, the museum provokes poets to confront the local effects of oppositions that have preoccupied discussions of poetry for the past 50 years: attitudes of homage toward the artistic forms of the past, and the impetus to innovate; the modernist isolation of the autotelic art object, and the postmodern destabilization of the terms of that isolation; the subjective passions of the singular observer, and the web of social and political tensions that ensnare and fracture individual perspectives. The poets I discuss in this study are not content to let these matters rest by choosing sides. As they notice the ways that a museum frames their particular encounters with art, a gallery visit becomes a space for interrogation, a circumscribed occasion for addressing one or more of the terms of these debates. Ekphrasis becomes a critical strategy by which poets negotiate the impasses that these oppositions present, and propose mediating positions that move beyond them.

The institution of the modern art museum focuses these oppositions with particular force because it embodies their convergence. It is the place where artistic innovation becomes tradition in a perennial cycle of vanguard effrontery and cultural enshrinement. It removes a work of art from its historical or creative context, holding it up for aesthetic contemplation on a pedestal or in the "white cube,"[6] even as it surrounds that work with governing narratives of its cultural environment and significance, and creates a context of its own—one for which artists may intend their work. Versions of these two tensions—tradition versus innovation, aesthetic appreciation versus political, social, or economic context—have preoccupied critics of the museum since its rise in the late eighteenth century.[7] Quatremère de Quincy, writing in 1796, opposed the creation of museums because they removed artworks from the religious and civic sites that gave those works moral purpose (Sherman 128–9). He recommended the city of Rome, "immovable in its totality," over the galleries of a public museum, which are "sterile and cold" (127, 128). More than a century later, in 1923, Paul

Valéry felt the same chill, lamenting the "cold confusion" (202) of a vast array of decontextualized artworks in a cultural shrine:

> A strangely organized disorder opens up before me in silence. I am smitten with a sacred horror. My pace grows reverent. My voice alters, to a pitch slightly higher than in church, to a tone rather less strong than that of every day. Presently I lose all sense of why I have intruded into this wax-floored solitude, savoring of temple and drawing room, of cemetery and school [. . .]. (203)

Valéry objects to the ways museums dampen the impact of innovative achievements (the presentation en masse of "rarities whose creators wanted each one to be unique" [204]), and he also objects to the ways museums ritualize and regulate responses to the works they apotheosize. Adorno, commenting on Valéry, underscores these objections and extends them to a critique of commodification:

> Museum and mausoleum are connected by more than phonetic association. Museums are like the family sepulchres of works of art. They testify to the neutralization of culture. Art treasures are hoarded in them, and their market value leaves no room for the pleasure of looking at them. Nevertheless, that pleasure is dependent on the existence of museums. (175)

Like these critics, the poets I address find the museum to be a source of anxiety and suspect pleasure, though their responses are usually less morbid. They ruminate on the transformative effects that museum-space has on art and audience, for good and ill, and in so doing they situate works of art in the exposed mechanisms of their decontextualization and their canonization.

Recent studies of the museum have sought to escape the conclusion that it always has a deadening, isolating, or aggrandizing effect on its contents: as Andreas Huyssen usefully observes, "We need to move beyond various forms of the old museum critique which are surprisingly homogenous in their attack on ossification, reification, and cultural hegemony [. . .]" (18). Huyssen calls for a re-examination of the museum that recognizes the ways it is a site of tension and confrontation:

> Few who have written on the museum [. . .] have thus argued that we need to rethink (and not just out of a desire to deconstruct) the museum beyond the binary parameters of avant-garde versus tradition,

museum versus modernity (or postmodernity), transgression versus co-option, left cultural politics versus neoconservatism. [. . .] I would suggest that the avant-garde's museumphobia, its collapsing of the museal project with mummification and necrophilia is one such claim that belongs itself in the museum. (18)

Criticism of contemporary poetry could also benefit from this kind of re-examination. A similar string of binary oppositions, acknowledged to be reductive and partisan yet relentlessly rehearsed, has dogged criticism of American poetry since the fifties. In their various guises, these binaries function as critical devices, periodizing categories, anthologizing principles, and shibboleths of affiliation: academic versus beat, cooked versus raw, mainstream versus avant-garde, poetry of accommodation versus poetry of opposition, classic versus outlaw, the "canon to the right of us" versus the "canon to the left of us."[8] The "anthology war" between Donald Hall's *New Poets of England and America* (1957) and Donald Allen's *The New American Poetry* (1960) was waged over these divisions, as were later battles between poetic positions represented by the "mainstream" Associated Writing Programs on one hand, and the "avant-garde" Language writers and their followers, on the other. Moreover, these divisions usually fall into predictable alignment under the two mammoth categories for stylistic, periodizing, and theoretical difference that have obsessed critics in the past three decades: modernist versus postmodernist. Self versus social space. Organic form versus the open field. Confession versus disjuncture. Lyric versus language. The terms change, but the two-column scaffolding remains. Jed Rasula observed (in 1996) that "the famous battle of the anthologies of 1960 [. . .] resulted in a stalemate that continues to mark the world of American poetry as we near the century's end" (138). Despite the consensus (in 2005) that these oppositions are exhausted and constrictive, as significant as the differences among groups of poets were and continue to be, they remain well-worn grooves in critical memory, sources of unresolved anxiety. The museum setting, I will show, offers poets the benefit of distance from their immediate sphere, and in so doing brings several of the terms of these entrenched dichotomies under review.

Rasula's study of canons in later twentieth-century American poetry is titled *The American Poetry Wax Museum*. The museum he speaks of is not an actual place, but a critical figure, an extended analogy for the mainstream poetry establishment, what Charles Bernstein calls "official verse culture" (6).[9] In making this comparison, Rasula overlooks the real presence of museums in postwar poetry—the poets have gotten there ahead of him, and they come from both sides of the anthology wars. Here is the end of Adrienne Rich's "Love in the Museum," in Donald Hall's anthology:

Or let me think I pause beside a door
And see you in a bodice by Vermeer,
Where light falls quartered on the polished floor
And rims the line of water tilting clear
Out of an earthen pitcher as you pour.

But art requires a distance: let me be
Always the connoisseur of your perfection.
Stay where the spaces of the gallery
Flow calm between your pose and my inspection,
Lest one imperfect gesture make demands
As troubling as the touch of human hands. (Hall 272)[10]

Here is the beginning of John Wieners's "A poem for museum goers," in Donald Allen's anthology:

I walk down a long
passageway with a
red door waiting for me.
It is Edward Munch.

Turn right turn
right. And I see my
                    sister
hanging on the wall,
heavy breasts and hair

Tied to a tree in the garden
with the full moon
are the ladies of the street.
Whipped for whoring. (Allen, *The New American Poetry* 371)[11]

These are strikingly different poems, to be sure. Rich's rhymes and iambic pentameter convey the cool restraint and formal perfection characteristic of fifties academic verse, a poetic mode that she announces: "art requires a distance." Wieners's irregular lines, strong enjambment, and pliable syntax reflect the experimental "composition by field" strategies of Charles Olson, with whom Wieners studied at Black Mountain College. With its sinuous lines and measured description, Rich's poem is calm and eloquent. With its percussive fragments and heavy alliteration, Wieners's is harsh and commanding. Yet the two poems share a setting and a stance: both speakers are self-consciously positioned in a museum, moving toward a "door," a point of entry into a work of art. Both name a particular artist whose vision dictates their tone and imagery: an approach to Vermeer requires crystalline detail, while an approach to Munch requires violent garishness. The ekphrastic encounter prompts both poets to contemplate a sensual exchange from a perspective of indeterminate gender; both examine an image of a woman's body and consider the troubling consequences of "the touch of human hands." Wieners's art, it turns out, also requires a "distance"—the dropped line and white space separate the speaker from a woman who is "sister," not lover.[12] For both poets, "the spaces of the gallery" invite meditations on specularity and desire. The museum, as a topical presence in twentieth-century poetry, crosses factional lines and represents a powerful point of intersection between opposing poetic positions.

But "always the connoisseur of your perfection"? Ekphrastic stances like Rich's in "Love in the Museum" make charges of elitism and escapism easy to level. The high arts, facing a shared "anxiety of obsolescence," huddle together in the museum for warmth,[13] ignoring the political and economic conditions that create and surround that enclave. In this view, ekphrasis is a form of cultural capital, of literary cachet—a sanctioned path to literary legitimacy via sanctioned works. One might read Hecht's "At the Frick" as announcing its location in the esteemed collection of the steel-industry millionaire, only to exclude all other considerations of that site from its purview. Hecht is concerned with what "Master Bellini" has made of St. Francis in Ecstasy and the craggy landscape around him, not with robber barons or Pinkerton guards (259).[14] These exclusions are appropriate to Hecht's subject—in such ecstasy, one surmises, the cares of the world slip away—but the proper noun in the poem's title must be sanitized of its worldly (as opposed to art-worldly) history for those cares to be forgotten. A poem similarly troubling in its vision of connoisseurship is Jorie Graham's "San Sepolcro," where the speaker establishes a privileged vantage point before addressing Piero della Francesca's Madonna: "I can take you there, / [. . .] / [. . .] This / is my house, / / my section of Etruscan / wall, my neighbor's / lemontrees, and, just below / the lower church, / the airplane factory" (21). The speaker begins from a position *above* the factory and the tourists, as a resident with special access to works of art: "[. . .] No one / has risen yet / to the museums, to the assembly / line—bodies / and wings—to the open air / market" (21–2). Like Hecht, Graham is concerned with an artwork's vision of transcendence in a space that is vacant of economic activities, though for her these peripheral matters do encroach on the view.

My focus in this study is on those ekphrastic poems that negotiate a path between visions of ecstasy and inquiry into the cultural conditions that foster (or counter) such visions. Ekphrasis, because it spotlights aesthetic perception—attention to art as art—, tends to be associated with the former poetic trajectory more than the latter, an assumption I hope to correct. Part of the problem with arguing for an expanded critical role for ekphrasis, or for poetry more generally, is that "aesthetic" remains an uneasy term in current critical discourse, one that provokes varying degrees of suspicion. As Denis Donoghue observes, the later decades of the twentieth century were difficult ones for proponents of aesthetic reading: "negative associations hung upon the words 'aesthete' and 'aestheticism' and upon the phrase 'art for art's sake.' 'Decadence' was just around the corner" (12). The larger institutional developments that contributed to these negative associations, including historicizations of the aesthetic as an unstable category, renewed Marxist critiques of aesthetic ideology, reactions against New Critical reading practices, canon reforms that eschewed formalism in favor of multiculturalism, and, in poetry in particular, the radicalization of form advanced by the Language poets, often reflected a common understanding of "aesthetic" as meaning the kind of connoisseurship and appreciative distance that poems like Rich's and Hecht's exemplify. Though "aesthetic" is less often thought of so narrowly today—few would reject aesthetic energy or pleasure out of hand, or claim that it is incompatible with critique—it remains the case that poetry is marginalized in English departments in part because it inherits the effects of this anti-aesthetic bias. Museum-conscious ekphrasis requires us to broach this subject, and to consider the ways that poets themselves have contended with it.

A few words about what this study is not. First, it is not a history of twentieth-century poetic canon formation, but a re-examination, through a particular lens, of some of the terms that underlie those formations. I do not resume arguments about canons and countercanons, or dwell on histories of groups and movements, because I feel the deck is in need of reshuffling: the so-called mainstream is more heterogeneous, and the influence of avant-gardes more pervasive, than most critics have allowed. By the end of the twentieth century, the American poetic field is not dominated by confessional lyrics, scenic interludes, and small epiphanies, but by an extremely wide variety, even cacophony, of efforts to experiment with and synthesize new poetic possibilities. Innovation as a poetic imperative is far

from marginal: as John Ashbery puts it, "To the 'newness' then, all sub-
scribe, albeit with a few reservations" (*Flow Chart* 159). Second, this
study is not an effort to theorize the category of the aesthetic. What aesthetic perception *is* varies from time to time and place to place, and even
in this limited selection of American poets writing about art in the late
twentieth century, attitudes toward aesthetic pleasure differ considerably
(what pleases Richard Howard is not what pleases Kathleen Fraser, for
example.) I am concerned with the circumstances and effects of particular
aesthetic moments as the poets define them. Several other related subjects,
though they come up in passing, are beyond the scope of this study:
poetic collaborations with visual artists, ekphrases of film or theatrical
images, the impact of reproductions and electronic media on art perception, and poets' engagements with the setting of the studio. I also do not
investigate poets' interactions with other types of museums, including
natural history museums, ethnographic and scientific exhibits, and folk
museums, but limit my inquiry primarily to poets' encounters with fine
arts museums and galleries, because they spotlight the aesthetic questions
with which I am concerned.

   This study is an intervention, through a series of case studies, into
problems of aesthetic perception and institutional critique in contemporary American poetry. The framework for discussion is interdisciplinary,
dovetailing discussion of contemporary poetics with museum studies. The
method is extrapolative and heuristic: I cull my examples from the vast
numbers of ekphrastic poems written in the past fifty years to examine the
patterns of description, interpretation, and critique that emerge when the
museum setting comes to the fore. I analyze these patterns by indicating
the critical mechanisms already at work in the poems, by reading them as
spaces for interrogation. In the process, I show how contemporary
ekphrastic poetry, in a variety of forms and arising from different poetic
traditions, enfolds several contemporary critical concerns. My method is
thus investigative but not entirely disinterested: I too strive to find a critical stance that does not disavow the aesthetic. I broadly apply Pierre Bourdieu's principle for elucidating the "differential stances" that a literary work
can exhibit, considering a poem as both "structured" and "structuring"—
that is, as fully explainable neither by extrinsic determinisms nor by a
purely internal subjectivism (Johnson 14). Ekphrastic poems, I claim, are
best read in this way. On one hand, they reflect poets' "competence in a
process of appropriation [. . .] [which] is a form of cultural capital" (23),
a competence that is predicated on institutional, educational, and social
determinants. On the other, ekphrasis looks out, and around: it reflects

poets' efforts to *mediate* that process of cultural appropriation—to call attention to its media and means—from within poetic acts of creation and critique.

Tensions between creation and critique are particularly heightened in ekphrastic poems that are written under the auspices of the museum itself. In chapter one, I examine the ambivalent attention to the museum we find in poems in museum-sponsored anthologies. In these volumes published by national and university museums, poets are acutely aware of the institution that shapes their experiences of works of art, and, in most cases, commissioned their poems. By way of introduction to a central tension in contemporary ekphrastic poetry between aesthetic liminality and critical interrogation, I analyze the competing orthodoxies these collections house, focusing on the texts and paratexts of four anthologies published in the past two decades: the Tate Gallery's *With a Poet's Eye* (1986), the Art Institute of Chicago's *Transforming Vision* (1994), the University of Michigan Museum of Art's *A Visit to the Gallery* (1997), and the Yale University Art Gallery's *Words for Images* (2001). These collections rely on a rhapsodic and idealized view of aesthetic perception that Bourdieu usefully summarizes as the "charismatic ideology"—the belief that we experience art as a descent of grace, an immediate and intuitive moment of rapture. But they are also peppered with moments of resistance to and critique of this ideology, skeptical inquiries that are no less predictable given the academic climate from which they arise. Examining three ekphrastic poems as case studies, I argue that the strength of these poems depends upon their ability to mediate these two positions and deploy their critical interrogations through attentiveness to artistic media and aesthetic pleasure.

In chapter two, I turn to the contemporary poet for whom this strategy, this effort to mediate the demands of aesthetic rapture and critical resistance, is part of his fundamental donnée—John Ashbery. It is telling that Ashbery, the only contemporary American poet who has been fully embraced by proponents of the lyric tradition and the avant-garde alike, has written the most famous ekphrastic poem of the latter half of the twentieth century, "Self-Portrait in a Convex Mirror." To dispel the critical myth that this dual canonization is possible because there are "two John Ashberys," one traditional and one avant-garde, I argue that the ekphrastic occasion focuses Ashbery's ambivalence about the opposition between tradition and innovation itself. Revisiting the museum scene of his avant-garde beginnings, I argue that from the outset the idea of the avant-garde, for Ashbery, is inseparable from its institutionalization. I then turn to the uneasy confluence of attitudes of homage and resistance in "Self-Portrait in a Convex Mirror," a poem where the problems of the museum setting—its insularity ("You can't live there"), its sanctity ("investing / Aura"), and its elitism ("the leisure to / Indulge stately pastimes")—are interrogated (*Self-Portrait* 79).

Museum poems by Kenneth Koch and Richard Howard, poets of the same generation who are associated with opposing coteries, are the subject of chapter three. Both Koch and Howard have well-honed peripheral vision, and both approach the setting of the museum with the same attitude—comic irreverence. I show how their museum comedies, humorous stagings of encounters with art and its institutions, complicate our understanding of the oppositions that are thought to divide these poets—tradition and contemporaneity, literary and popular culture, modernism and postmodernism. Koch's career-spanning objection to artistic conformity extends to a critique of the conventionalization of experiment itself, the obligatory effort to trump the latest round of making it new. Museum settings, because they spotlight the consecration of innovation, are fertile ground for parody, for an interrogative stance that never loses sight of aesthetic pleasure. Claimed by critics as an exemplary postmodernist, Koch devises fabular museums in which to parody avant-garde innovation and the postmodern apotheosis of inclusiveness. Similarly, Howard's preoccupations with the institutional "trappings" of aesthetic experience run counter to the usual terms of his reception as a traditional poet. His museum poems, elaborate impersonations and ekphrastic forays, self-reflexively ironize his participation in and conservation of the literary traditions of his sources. I argue that Howard's "atavistic postmodernism" is a strategy of critique that preserves the curiosity and brio of an aesthetic approach to the museum's "omnivorous package."

In chapter four, I turn to the recent work of three poets whose reliance on the "museum as muse" is complicated by awareness of the gendered terms of ekphrastic looking: the muse, as Anne Carson explains, is a woman with "a face / and a past / worth painting" (*Plainwater* 50). Examining ekphrastic poems by Carson, Cole Swensen, and Kathleen Fraser, I argue that the museum-consciousness we find in them challenges another reductive critical opposition, the application of gender difference to the gazes and objects of an ekphrastic exchange. As a corrective to critical overemphasis on antagonisms between male gazes and feminine objects, I propose instead to read these poets' acute awareness of the museum setting as a form of poetic site specificity—strenuous engagements with physical, institutional, and historical places. I argue further that these poets exercise this site specificity through an amplification of the linguistic mode of *deixis,* a mode that enables them to interrogate the "here and now" of their ekphrastic utterances. Deictic site specificity allows these poets to find a critical stance attuned to particular subjective pleasures and longings. In

Swensen's choreography of the perceptions of a defiant museumgoer, Fraser's revisionary reading of a canonical masterpiece, and Carson's portrait of a recalcitrant muse, we find feminist critiques saturated in aesthetic awareness.

For the poets I discuss, the museum is not only muse but stimulus for critical insight. The title of the MoMA show with which I began this introduction tells only half the story: it smoothes the complexities of artists' responses to the museum under a heading of inspiration—the museum is finally a source of creativity as opposed to ambivalence, confusion, or conflict. The poets I address, like their artist counterparts, are open to all of these possibilities, and reluctant to give up any of them. As they frame ekphrasis in the occasion of museumgoing, they find compelling ways to navigate the tensions the setting brings to the fore. Taken collectively, their approach is most analogous to Struth's penetrating eye for the dynamics of the museum situation. In his photograph of museumgoers looking at Pollock's painting, Struth frames reactions to innovation. Another photograph, of museumgoers in front of Veronese's enormous "Feast in the House of the Levi" in the Galleria dell'Accademia in Venice, might be read as a contrast— a glimpse of reactions to tradition. But the two photographs have the effect of drawing the experiences of seeing Pollock and Veronese together—both give us works of overwhelming scale and intensity considered scandalous in their day, and both provoke intent postures and restless motion. Struth calls our attention to the apparatuses that shape our experiences of art, whether in the Venice Academy or MoMA—the Corinthian columns that extend the imagery of Veronese's painting into the dimensions of the room, the white wall that isolates the Pollock, the metal rails that keep visitors from getting too close—even as he reminds us of the power of the artworks themselves. The poets in this study are engaged in similarly attentive negotiations of art and its environments. Contemporary ekphrastic poems that foreground and interrogate the museum setting comprise a set of particularly lucid engagements with the defining tensions between aesthetic pleasure and institutional critique, cultural enshrinement and artistic innovation, in postwar American poetry. The mode of vision, critique, and mediation we find in them is a vital and discerning avenue in contemporary poetic practice.

Chapter One
Charisma and Critique: Interrogating Aesthetic Liminality in Museum-Sponsored Anthologies

The first poem in *Transforming Vision* (1994), a collection of writers' responses to works in the Art Institute of Chicago, sets a tone of meditative escape:[1]

> Today as the news from Selma and Saigon
> poisons the air like fallout,
>   I come again to see
> the serene great picture that I love.
>
> Here space and time exist in light
> the eye like the eye of faith believes.
>   The seen, the known
> dissolve in iridescence, become
> illusive flesh of light
>   that was not, was, forever is.
>
> O light beheld as through refracting tears.
> Here is the aura of that world
>   each of us has lost.
> Here is the shadow of its joy. (*Transforming* 18)[2]

The poet is Robert Hayden, and the "serene great picture" he has come to see is Monet's *Water Lilies* (1906). The speaker enters the museum to forget the political upheaval that troubles his moment, to find sanctuary from news of racial violence and the Vietnam War. For Hayden, Monet's almost abstract vision of water and light evokes an eternal realm of transcendent harmony in which "The seen, the known / dissolve in iridescence [. . .]." The final stanza apostrophizes light in a lament for lost innocence, and suggests that this museum experience allows an "aura" of that lost world to be regained. Based in feelings of love and faith, the speaker's response to the painting exemplifies what Bourdieu has called the "charismatic ideology," the belief that aesthetic experience is a spontaneous "descent of grace (*charisma*)" (*Love of Art* 54)—a visitation, as the theological term suggests, of a divinely conferred gift or power. In a volume that gathers responses to art under a heading of "transforming vision," Hayden's poem is an appropriately visionary starting point.

Yet the frame of the anthology has promoted this reading of the poem as an escapist retreat into aesthetic solitude. The poem first appeared in Hayden's 1970 collection *Words in the Mourning Time,* a book that "respond[s] directly to one of the most violent decades in American history," and includes a poem on Malcolm X and the title poem in dedication to Martin Luther King, Jr. and Robert Kennedy (Oehlschlaeger 115). "Monet's 'Waterlilies,'" part of a section that contemplates an antithesis between political violence and artistic serenity (115), reflects the news of early 1965, when King's demonstrations for voting rights provoked violent reaction in Selma, Alabama, and U.S. involvement in Vietnam escalated, the government in Saigon under Nguyen Cao Ky rapidly deteriorating. In *Words in the Mourning Time,* the poem's first stanza had two additional lines describing the canvas as one that "flames disfigured once / and efficient evil may yet destroy," a reference to damage done to several of Monet's late paintings by Nazi artillery fire (55, Oehlschlaeger 116). Hayden cut these lines when the poem appeared in *Angle of Ascent* (1975), and the later version is reprinted in *Transforming Vision,* where it is dated, misleadingly, for its posthumous publication in his *Collected Poems* (1985). Depoliticized by Hayden's revisions and by its separation from the other poems in *Words in the Mourning Time,* the poem appears to be less engaged with the troubled relationship between art and politics than it does in its original context. Moreover, when we come upon its charismatic fervor in the anthology, it is easy to overlook the traces of political tension that remain. The nostalgic final quatrain mirrors the first, with its blunt naming of the reasons why innocence has been lost—Selma, Saigon, fallout. The fact that the speaker "come[s] again to see" Monet's painting indicates that the museum visit is habitual and temporary, a pause before re-entering the "poisoned air." Hayden suggests that we do not find serenity in a museum without being reminded why we need to search for it, and he gives us two causes for "refracting tears" in the form of precise geographical foci.

The museum thus provides a double frame for reading Hayden's ekphrasis: the speaker encounters Monet's painting in a museum, and a museum publishes the result under its aegis. In this case, the effect of that double frame is one of isolation and mystification, a drawing of a boundary between aesthetic experience and the political world. Though the details of the poem suggest the tenuousness of that boundary, it remains for the most part intact. Hayden's speaker aspires to what Carol Duncan calls a condition of "liminality": he crosses a threshold into a space apart, where he experiences "a state of 'detached, timeless and exalted' contemplation that 'grants us a kind of release from life's struggle [. . .]'" (12).[3] But few poets, and few museumgoers, have Hayden's faith. In the many museum anthologies published in the past two decades, four of which will be my focus in this chapter, Hayden's "liminal" experience represents one pole in a spectrum of responses to the museum setting. Poets in recent years have come to museums not only because they have sought illumination or solace, but because museums have invited them in, commissioning ekphrastic poems about works in their permanent collections. In the Tate Gallery's *With a Poet's Eye* (1986),[4] the Art Institute of Chicago's *Transforming Vision: Writers on Art* (1994), the University of Michigan Museum of Art's *A Visit to the Gallery* (1997), and the Yale University Art Gallery's *Words for Images: A Gallery of Poems* (2001), the majority of poems were written for the occasion.[5] That occasion, I will show, provides a valuable lens through which to read a persistent tension in contemporary poetry and in contemporary literary criticism more generally, a tension between competing orthodoxies I label "charisma" and "critique."[6] On one hand, Hayden's hand, we find approaches to aesthetic perception as sudden ravishment, as escape from or avoidance of political and social exigencies. On the other, we find that the museum frame provokes resistance and interrogation, skeptical renunciations of the liminal experience Hayden seeks, and exposures of its political and artistic dangers. One stance relies on disinterested aestheticism and ideas of modernist aesthetic autonomy; the other, on the imperatives of postmodern institutional critique.

Poetry anthologies have been a subject of critical scrutiny in recent years, but museum-sponsored volumes have been overlooked.[7] Most studies of poetry collections have focused on the ways they define opposing poetic canons, taking particular anthologies to task for partisan narrow-mindedness, and celebrating others for providing alternatives.[8] These studies have remained for the most part within the narrow limits of the poetry world, and have not considered how anthologies of poems in other contexts (and which place poems in the company of prose pieces and other art forms), reflect critical tensions broader (and other) than the antagonisms between mainstream and avant-garde camps in contemporary American poetry. Museum anthologies offer a different institutional perspective, one that intersects with a wider field of aesthetic and critical concerns, even as they offer a less than representative survey of poetic groups and practices. The poems in these anthologies represent the poetic mainstream, almost by definition: inclusion in these books requires an established reputation, a literary dossier, that would recommend a poet to a museum. That reputation may be established at a national level, as is the case with contributors to *Transforming Vision,* which includes Poet Laureates (Richard Wilbur and Rita Dove), Pulitzer Prize winners (Stanley Kunitz and Garry Wills), and canonical writers of the past (Willa Cather, Carl Sandburg, Wallace Stevens), or it may be established at a regional or institutional level, as is the case with contributors to *A Visit to the Gallery* and *Words for Images,* with their links to the University of Michigan and Yale University, respectively. Not surprisingly, given the reliance in these books on ekphrasis as a traditional genre, Language poets are not out in force here, and there is a dearth of formal experimentation—a museum-sponsored anthology, with its implicit homage to consecrated works of high art, and its traffic in cultural cachet, is on the whole a conservative proposition. That said, I aim here to put the poetic mainstream to a use it has not often been made to serve, tied up as it has been as the foil of the avant-garde: the uneasy admixture of attitudes we find in these anthologies represents not lyric homogeneity, but a range of inquiries into received notions of aesthetic liminality, artistic production, and cultural capital.[9] The ekphrastic assignment, because it displaces attention outside of poetry, prompts examinations of the assumptions, cautions, and commonplaces of the critical moment in which these poets write.

This chapter examines museum-sponsored anthologies in order to delineate the tensions between charisma and critique that arise in contemporary ekphrastic poems, and to adumbrate the issues surrounding ekphrasis and museums that will arise in the rest of this study. Identifying the relative propensities for charismatic appreciation and critical interrogation in individual poems in these collections, I follow each mode of response, or train of argument, into its pitfalls and contradictions. In the first section, I make the case that charisma is only half the story: these anthologies depend upon and provide myriad examples of idealized attitudes toward art perception, of the mystifying aesthetic longings that circulate readily in the museum atmosphere, but they also countermand those attitudes. In the second section, I argue that critiques of the museum, and skepticism about aesthetic dispositions, are no less predictable in a collection of late-twentieth-century writings from an academic milieu. I describe the shortcomings of some of the poems in these anthologies as symptomatic of an inability to reconcile the demands of critique with the pleasures of art. Finally, I offer close readings of three ekphrastic poems that negotiate this impasse. The most successful poems in these anthologies are those that mediate between charisma and critique, and their strength depends upon their ability to deploy their critiques through engagements with artistic media and aesthetic response.

THE RITUAL PRECINCT OF THE AESTHETIC CULT

Advice for reluctant or perplexed museumgoers often takes this familiar form: allow the art to speak to you. Don't worry about understanding it, just experience its serenity (or vivacity, or rage). Bourdieu, in his sociological analysis of the responses of middle-class European museumgoers in *The Love of Art* (1969), calls the set of principles that underwrites this advice the "charismatic ideology," the belief that art perception is spontaneous, intuitive, and passive, and the attendant belief that artworks are autonomous, luminous, and sanctified. These beliefs, Bourdieu argues, mask the social and educational conditions—the cultural capital—needed to set the stage for aesthetic responses.[10] In *The Field of Cultural Production* (1993), he describes three central features of this ideology. First, it "attributes to the work of art a magical power of conversion capable of awakening the potentialities latent in a few of the elect": aesthetic awareness comes as a gift, not a skill. This belief in turn "contrasts authentic experience of a work of art as an 'affection' of the heart or immediate enlightenment of the intuition with the laborious proceedings and cold comments of the intelligence": one does not have to study to receive such grace. Finally, a charismatic approach to art "treat[s] as a birthright the virtuosity acquired through long familiarization or through the exercises of a methodical training," and maintains "silence concerning the social prerequisites for the appropriation of culture" (234). Tied to the particular conventions of the museum setting, Bourdieu's term usefully encapsulates the rhapsodic tendency that poems about art may exhibit, the common thread of fervent homage and ritual zeal we find in these anthologies—faith in art that is blind to its mechanisms of production and reception. But I employ Bourdieu's concept here to complicate it and point out its limitations: museum-sponsored anthologies exhibit and rely on faith in charismatic aesthetic response, even as they document its instability.

On first inspection, these anthologies consolidate the values Bourdieu describes. First, they are founded on the premise that art can awaken "the potentialities latent in a few of the elect." As the title of the Tate anthology indicates, these books offer readers a chance to see "with a poet's eye," with a gifted sight that perceives artworks more freshly, emotionally, and imaginatively: Pat Adams's introduction tells us that "the poet, through a personal vision and expression, brings to the subject a wider, sometimes entirely unexpected dimension [. . .]" (*Poet's Eye* 8). The introductory statements in all of these books emphasize the special nature of their contributors' approaches to art. Museum director William J. Hennessey tells us that "*A Visit to the Gallery* unites the talents of twenty-four outstanding contemporary writers with masterpieces from the Museum's permanent collection" (*Visit* 9). He praises the editor "for serving as resident muse for the project and for recruiting a truly stellar group of contributors" (9). Similarly, *Transforming Vision*'s jacket copy stresses that exceptional writers are gathered here to "share their visions of the museum's masterpieces": "A perfect blend of visual and verbal eloquence, *Transforming Vision* will be meaningful for devotees of literature and art alike." These "paratexts" [11] are the customary congratulations and acknowledgments of a collaborative publishing venture, but in retaining the language of aesthetic devotion—muse, talent, masterpiece, stellar, vision— they present the books' contents as the products of inspired mediums.

These anthologies also rely on a hyperbolic rhetoric of the "magical power" of art and its environments. In the introduction to *Transforming Vision*, Edward Hirsch admires the ways writers "capture a sense of the museum's persistent and unsettling magic, the urgency of being pulled back by certain talismanic paintings" (9). Noting that the writers were "left to their own devices to choose—to be chosen by—any work that intrigued them" (10), he emphasizes art's attractive powers and the poets' freedom from the constraints of commentary. Similarly, Richard Tillinghast notes in the introduction to *A Visit to the Gallery* that "The variety of [poets'] responses pays tribute to the dazzling power of the imagination when it is set free to wander and discover. And what a place for wandering and discovery the University of Michigan Museum of Art turned out to be!" (12). He presents the museum as a place where poets "wander" freely, not a place where art-historical narratives or learned familiarity guide the way. Magic, not method. Hirsch is impressed when works are recalled with "rapturous exactitude" (*Transforming* 10), and he offers this quasi-mystical speculation: "It would seem the Art Institute was a secret each of us had discovered on his or her own. Many felt called upon to testify to what they had beheld" (10). Hirsch's introduction addresses other aspects of the ekphrastic assignment, including the importance of considering artworks by "situating them in history or in experience" (9), but the occasion calls for a heightened rhetoric of aesthetic isolation: "The pieces in *Transforming Vision* are filled with intimacies attained, with reflections, refractions, revelations" (11).

Even before they are opened, these elegant and expensive books[12] create a mood of charismatic appreciation and disinterested aestheticism. The jacket illustration for *Transforming Vision*, Matisse's *Woman Before an Aquarium* (1921), presents a woman watching goldfish with her chin on her hand, a small blank sheet of paper on the desk beside her. In the line of her gaze, as if seen through the fishbowl, is the book's subtitle, *Writers on Art*. The woman becomes a figure for the writers whose "transforming visions" will make up the collection, suggesting that the museum experiences related within, like this image, reflect a process of moody, leisurely, solitary contemplation of a contained, silent sphere. The cover of the Yale anthology evokes a similar mood of "liminality" as it depicts an actual threshold. Wrapping around the jacket is Edward Hopper's *Rooms by the Sea* (1951), in which the door of an empty room opens immediately onto what John Hollander describes as "an uncontainable expanse of ocean" (*Words* 72). The cover itself is a door-shaped frame: we are invited to step over the threshold of the domestic and into the sublime. As Loizeaux observes of other museum-produced books, packagings like these—the high-quality paper, the abundance of color reproductions, the ample white space surrounding both texts and images—create an atmosphere of high-cultural "ease and leisure," a place "where there is time and space for contemplation, an ordered world separated from the ordinary" (81).[13] It is also, she points out, "a world of money that can afford such time and space, and, of course, that can purchase the paintings and sculptures that furnish it" (81). The promoters of these books, however, emphasize contemplation over commerce (to employ Elizabeth Bishop's opposition).[14]

Even more so than the others, the University of Michigan anthology frames its contents in the insularity of museum space and the self-reflexivity of the aesthetic encounter. Its jacket illustration, Pier Celestino Gilardi's *A Visit to the Gallery* (1877), from which the anthology takes its title, makes the museum setting explicit. Three young women in Victorian dress gaze either at the Medici Venus (whom we see in rear view), or at a fourth young woman standing beside the statue, which in turn seems to gaze into the mirror above the sofa. By placing this riddling image of mirrored gazes on the cover, the editors call attention to the self-conscious nature of the book's project: here is a work of art in a museum that depicts museumgoers gazing at a work of art in a museum. The book's unusual design by Beth Keillor Hay and Margaret Ann

Re emphasizes this centripetal pull. The jacket is cut away in the upper right-hand corner to show the cover beneath it. When unfolded, it reveals a loose scalloping line that runs diagonally across the leaf pattern of the cover, and then across the pattern of irregular lozenges and trapezoids on the inside flaps. The leaf pattern, we realize, is an enlargement of the upholstery of the sofa in the painting, and the pattern of the inside flaps is a close-up view of the pink and green marble of the gallery floor. Both magnified patterns reveal the cracks in the painted surface, suggesting both an antique elegance, and the idea that opening this book invites scrutiny of the intricate internal patternings of works of art. Considered as a physical object, *A Visit to the Gallery* presents itself as a self-referential *objet d'art*.

These introductory images and texts prepare us for isolated moments of aesthetic ravishment, and indeed many passages in the poems that follow suggest faith in aesthetic perception "as an 'affection' of the heart or immediate enlightenment of the intuition" (Bourdieu, *Field* 234). Glimpses of iridescence and aspirations for transcendence like we find in Hayden's poem abound, usually recounted in dramatic present tense as episodes of arrested perception and high romantic feeling. Rachel Wetzsteon's "Spring (The Procession)," after Joseph Stella's 1916 painting of that title, catches the speaker in a Shelleyan gust of nostalgic emotion:

> A thousand leaves rush forward:
> bright, like an image of something lost,
> quick, like a portent of something fast
> becoming a page in a tear-stained book [. . .]. (*Words* 12)

Karl Kirchwey's "Dialogue" with Giacometti's *Hands Holding the Void* (1934) approaches the enigmatic figure as if petitioning a sibyl:

> *What is the slot-eyed head that sleeps*
> *on the angled bench, and slyly keeps*
> *a ruminant's tongue for prophesy?*
> —It dreams of the pure idolatry
> of imagination, its consummate solitude. (*Words* 38)[15]

These anthologies are full of moments of "pure idolatry of imagination" and "consummate solitude," as here in Charles Baxter's "A Disappearance," after Whistler's *Sea and Rain* (1865):

> A human being on the point of vanishing,
> stepping forward and absorbed by air and by the tow
> of water flowing inward, Whistler's small man [. . .]. (*Visit* 39)

Note the mimetic enjambment on "tow": Baxter dramatizes a moment of dissolution in which a "small man" loses himself in an elemental power. This longing for a horizon of transcendence, this effort to distill perception and human concerns into one iconic image, is presented most concisely in Ken Mikolowski's poem after Franz Kline's *To Win* (1961). I quote the poem in its entirety:

> that
> single moment
> the truth
> of the thing (*Visit* 61)

Mikolowski's is the extreme case, and one hopes it is in some measure ironic—Kline's painting, dominated by a black trestle but splashed with red and brown, is not reducible to a single moment or truth. Yet ekphrastic moments like this one, in which the poet attempts to freeze the encounter with a work of art into a single moment of clarity or vision, appear throughout these books.

This pitch of sublimity and solitude can hardly be sustained: even these poems with their visionary glimpses are crossed with other agendas. Wetzsteon has already reminded us that the bright, portentous moment becomes, problematically, a text. The omen diminishes to a token of wistful revisionary memory: "a page in a tear-stained book / that people look at in separate rooms, thinking / *There was a thing called spring, and it gave / my better days a meaning* [. . .]" (*Words* 12). Kirchwey's "Dialogue" goes on to ask questions about the other kinds of extremity suggested by the figure's gesture and expression—bodily ("metastasis"), biblical ("Capernaum"), and historical ("Argonne," "eight million dead") (*Words* 38). Baxter views the erasure of "Whistler's small man" in a national context of tensions between "the British style" and "that American, that ire apostle" (*Visit* 39). Charismatic fervor appears in all of these poems, but it is accompanied and interrupted by argument, worry, idle chatter, distraction. Peter Sacks, in a poem in the Yale anthology about a carved ivory head from Congo, detours in the middle of an intense meditation on the head's inscrutable power to remind us that any reverie in the presence of an art object is framed: one must "[. . .] see it here, the place reserved for craft, for prayer, [. . .]" (*Words* 58). "Here" is the museum, the obligatory site of such encounters with prized artifacts. It is a place that has been "reserved" (in the double sense of 'set aside' and 'restrained') for two particular kinds of attention: appreciation of formal perfection or "craft," and a reverential attitude of "prayer." But the carved head allows no interlude of calm "re-imagining" once the speaker becomes aware that it is a racialized "monument" to a continent in a museum context. Sacks's reminder is telling: as soon as we notice the museum frame, the charismatic spell is broken.

Many theories of museum space, including Bourdieu's and Duncan's, fail to acknowledge that this countertendency, this spell-breaking rather than spell-binding effect, often coexists with idealized attitudes toward art—the museum setting as disturbance, not source, of aesthetic ecstasy. Critiques of the museum, as I will discuss further in the next section, often rely heavily on consolidating an "aesthetic" viewpoint in negative opposition to the critics' own position—and on assuming that most museumgoers actually swallow the "charismatic ideology" whole. When she traces the history of the concept of "liminality" as foundational to the modern museum, Duncan describes the continued sway of Kantian aesthetic idealism in terms that imply the existence of total believers:

> In philosophy, liminality became specified as the aesthetic experience, a
> moment of moral and rational disengagement that leads to or produces
> some kind of revelation or transformation. Meanwhile, the appearance
> of art galleries and museums gave the aesthetic cult its own ritual
> precinct. (14)

Duncan's language indicates clearly that she is skeptical of experiences that offer "some kind of" revelation—a "transforming vision," in this view, smacks of delusion. The word "cult" suggests debased zealotry and equates "aesthetic experience" with the behavior of fanatics. Duncan is concerned to show that the museum scripts a ritual scenario, and her analysis depends on a substantial congregation of aesthetic enthusiasts: despite her concession that "[in] reality, people continually 'misread' or scramble or resist the museum's cues" (13), she needs to emphasize that "it is the visitors who enact the ritual" (12). Similarly, Bourdieu's sociological position requires a predictably responsive population of aesthetes, and can reflect the congealment of an oppositional posture into an anti-aesthetic boast:

> Sociology and art do not make good bedfellows. That's the fault of art
> and artists, who are allergic to everything that offends the idea they have

of themselves: the universe of art is a universe of belief, belief in gifts, in the uniqueness of the uncreated creator, and the intrusion of the sociologist, who seeks to understand, explain, account for what he finds, is a source of scandal. (*Sociology in Question*, qtd. in Dunn 87–8)

The notion that "artists" are a univocally cultish bunch who are "allergic to everything that offends the idea they have of themselves" is a convenient critical fiction, but an exaggeration at best. Not only are many artists and poets unfazed by sociologists, but many share with them a desire to "understand, explain, [and] account for" what they perceive. If we look again at museum-sponsored anthologies, collections of writings that are inextricably entangled in a "universe of belief, belief in gifts," as we have seen, we find that the museum frame reveals fault-lines of internal contradiction that destabilize the charismatic universe of beliefs from the outset.

Bourdieu's third point about the "charismatic ideology," the requirement that the educational preconditions of the experience of art be ignored in a haze of rapture, does not hold true in these anthologies. The artworks that are presented, both poems and images, are not autonomous at all, but surrounded by explanatory texts, from the tables of contents to the footnotes. The editors may insist on readerly and writerly freedom to discover, but several paratextual layers steer the encounter. The introductions and prefaces contextualize the works in art-historical and literary commentary, and the pages that conclude the volumes, like the last rooms of exhibitions with their catalogs and reference books, remind us that we should consult the authorities. In *A Visit to the Gallery*, Ellen Plummer's endnotes offer brief interpretations of the paintings, information about their acquisition, and some art-historical context. In *Words for Images*, each entry is followed by a brief essay by Joanna Weber about the art, and another essay by Hollander about the poem. *Transforming Vision* ends with a "Checklist" (this is the term that heads the page) of the works in the museum's permanent collection that the writers address, each item followed by the work's size and donor. The book thus functions as a workbook for thorough exposure to the museum's great works, an educational blueprint for fulfilling a cultural obligation. Even as encounters between poets and artworks are touted as elect visions, they appear in a particular institutional context whose markers circulate throughout.

Moreover, the social and educational determinants of these encounters with art are not veiled but advertised. The Yale anthology is the most flagrant: the first criterion for inclusion in this collection is a Yale degree. As Hollander writes in the introduction, "Each of the poems collected here was written, specifically for this volume, by a poet with a degree from Yale, in connection with a particular object of twentieth-century art in the Yale Art Gallery" (*Words* xv). The anthology does not hide its transactions of cultural capital; it offers an annual report. As part of the celebration of Yale's tercentennial, the book is designed to "bring poets who were once students back to the Yale campus" (ix), a homecoming aimed as much at alumni patronage as inspiration. Serving this function, the book announces its educational preconditions as central to the capacities of the poets themselves. Hollander writes:

> For one thing, many poets of the later twentieth century have had con-
> siderable exposure to art-historical teaching and writing: they notice
> things—albeit as poets—that scholarship has taught them to see almost
> as inhering in the work of art itself. For another, they have behind (or
> would it be better to say, 'within') them a considerable history of
> ecphrastic poetry, from Homer until the present day. (xv)

Hollander acknowledges that "scholarship has taught them to see" aspects of the art that only *seem* to inhere in the works themselves. But even so, his two asides attest to the force of charisma: these writers notice things with a mark of election "as poets," and they absorb and carry their educations "within" the interior of poetic subjectivity. The other anthologies are less explicit about Ivy League privilege, but their contributors' notes likewise testify to the educational and professional credentials (the degrees, prizes, appoint-ments, and publications) of most of the participants. There are no solitary geniuses or "uncreated creators" here, but groups of writers who depend on safety in numbers and on the terms of high-cultural achievement.

So far Bourdieu and I might reach the same conclusion about these anthologies as artifacts in the field of cultural production: they present aes-thetic experiences that are predicated on and productive of social distinction. My point is, first, that they make no pretense about that fact—*announce-ment,* not suppression, of elite cultural requirements is necessary to the proj-ect—and further, that awareness of the institutional frame introduces an important element of questioning into poets' approaches to aesthetic percep-tion. The Tate anthology is a telling case—the museum itself, as national institution and beloved cultural repository, is front and center. We are repeat-edly made aware that the museum has solicited these poems, and many of the participants feel a need to acknowledge their debt. Gavin Ewart offers a "Magical Mystery Tour" that nostalgically recalls school field trips to the Tate's galleries. Colin Archer makes Blake's Nebuchadnezzar into an unlikely museum guide to key masterpieces. Elizabeth Jennings's "The Tate Gallery," which opens the volume accompanied by a photograph of the museum, invites readers to enter the building with an open mind, due respect, and compliance with security procedures. Mention of bag-checks for firearms and bombs leads to a cheerful warning against the artworks' figurative explosions. Jennings then cautions museumgoers about the assaults of international artists (Rothko, Ernst, Magritte, and Modigliani), pointing them instead with national pride to Blake. Finally, she imagines an after-hours visit to the museum in which perception is heightened and "The Tate's pure purpose" is certain (*Poet's Eye* 16). Jennings draws attention to the museum in order to endorse it, creating a tribute that notices institutional frameworks but elides their significance in an assertion of the purity of aesthetic purpose.

When it cohabits with awareness of institutional framing conditions, however, the pure purpose of charismatic reverence tends to break down. Even amidst congratulation and endorsement, we find moments of critique. Several poets in the Tate anthology are concerned about the museum as a falsely sanctified space, a place where a restrained atmosphere does a disservice to the art itself. John Wain describes how a Turner painting seems to "shriek" in a hushed room where "poised appreciations" are expected (*Poet's Eye* 40). Museum etiquette requires a lowered voice and polite behavior—as Valéry put it, the voice must alter "to a pitch slightly higher than in church, to a tone rather less strong than that of every day" (203)—but the artwork, Wain observes, is angrier than its setting allows. In "The Rothko Room," Gillian Clarke makes a similar point about "Turner's turbulence" and then shifts her focus to the periphery of the Rothko gallery, where an Indian guard falls asleep, "marooned" in the painting's blacks and reds (*Poet's Eye* 116). Prayerfulness is disrupted when the speaker notices the way the guard is trapped in the geometry and color of the artwork nearby. Naming the ethnicity of this museum employee, the observation frames an experience of the abstract work in a national context with complicated, if so far unexplored, implications.

Throughout the Tate anthology, the conditions of the museum experience attract notice, and not always uncritically. Ruth Silcock addresses the economics of museumgoing by describing Rodin's *The Kiss* as a work in which the public owns stock, a work which has been literally purchased by museum traffic (*Poet's Eye* 72). She points out the museum's democratic objective and holds it up for scrutiny, imagining ownership of "his right thumb-nail" or "her ankle bone," and observing that insurance and transportation costs underlie the sculpture's sensual presence (72). Silcock gently ironizes the notion that we all "own" art by reminding us that the public pays for its romantic cultural inheritance (and its dominant gender stereotypes) through a bureaucratized and ritualized system of appreciation.[16] Elizabeth

Bartlett also examines the museum's economic and political foundations in "Millbank," a poem that explores the implications of the Tate's location on the site of the former Millbank Penitentiary (1812-1892). For Bartlett, noticing the incommensurability of the two sites makes the museum experience physically and intellectually uncomfortable—the speaker and her companion jostle themselves trying to enter through a single opening of a revolving door. The speaker then imagines that she walks over the faces of the dead in her high heels (*Poet's Eye* 79). There is little room here for rapture or spontaneous grace: the past is haunting, and the present provokes a self-conscious awareness of an uncertain inheritance of guilt. On leaving the museum, Bartlett observes, the wind is "cold as an iron bracelet" (80).

In the critical gestures of these last examples, uncomfortable awareness of the ways the museum shapes encounters with art dispels the myth of immediate and intuitive aesthetic perception. Undertaking the assignment of writing an ekphrastic poem in a museum setting requires "laborious proceedings and cold comments of the intelligence"—selection and exclusion, interpretive distance, attention to detail, critical appraisal. Throughout these anthologies, the proceedings of the ekphrastic enterprise prompt responses that operate in very different registers than the "charismatic." David Slavitt, irreverently, calls one of Jackson Pollock's paintings "a big mother"—it requires the museum to have a large display space as well as large financial resources (*Words* 68). John Burt calls Malevich "exhibit A of Bourgeois Art" (*Words* 9). Sylvia Kantaris steps back from a painting of girls in a church choir with a critical observation: awareness of gender gives the picture "a double edge of satire" (*Poet's Eye* 68). This critical register and its pitfalls is the subject of the next section, but the point I want to make here, contra Bourdieu, is that charisma is usually adulterated by it. In a discussion with Hans Haacke published as *Free Exchange* (1995), Bourdieu approves of art that takes a critical turn:

> What strikes me about your artistic approach is that your work as a critical artist is accompanied by a critical analysis of the art world and of the very conditions of artistic production. The two forms of investigation nourish each other: your quasi-sociological observations and reflections are fully integrated into your artistic work. (1)

Haacke's critiques of corporate patronage comprise particular efforts to expose and resist the mechanisms of institutional control,[17] but he is not the only one to integrate artistic goals and "quasi-sociological observations and reflections": we find critical attention to the "conditions of artistic production" in the very seat of charismatic desire.

SPECIOUS UPLIFT AND OTHER EXTRANEOUS REASONS: CRITIQUE AND ITS COMMONPLACES

In the academic climate in which these anthologies were published and most of the poems written, critiques of aesthetic liminality are as familiar as the idealized and meditative aesthetic overtures they seek to challenge. The first response that "museum-sponsored" is likely to elicit is one of suspicion. Douglas Crimp, in *On the Museum's Ruins* (1993), observes that "the modern epistemology of art is a function of art's seclusion in the museum, where art was made to appear autonomous, alienated, something apart, referring only to its own internal history and dynamics" (13). For Crimp, the museum is a "figure" (his term, 13) for the discursive system that produces modernist aesthetic idealism: it preserves art's lofty status, isolates art from everyday concerns, and shores up distinctions between high and popular culture (21, 248). His aim is to theorize a shift from modernist aesthetic autonomy to a postmodernist critique of the institutions that foster that idea of autonomy, and in so doing, he claims, "the target of my critique of the museum is the formalism that it appeared inevitably to impose on art by removing it from any social context" (25). Crimp articulates what has become a postmodern critical cliché of "resistance" to several common enemies:[18] formalist interpretations that would exclude a work's political, social, or material significance, elitism and class privilege, and the hegemonic control of artistic production and reception by institutions like museums.

Stressing the continued need to confront these problems (an emphasis I endorse), Crimp identifies examples of conservative retrenchment in the face of postmodern critique, efforts to reinstate ideals of aesthetic autonomy and to "reestablish the traditional fine arts categories by all conservative forces of society, from cultural bureaucracies to museum institutions, from corporate boardrooms to the marketplace for art. And this has been accomplished with the complicity of a new breed of entrepreneurial artists [. . .]" (272). Poets who write for museum-sponsored anthologies might be considered part of this breed. Like the visual artists whom Crimp criticizes here, these contributors can be charged with opportunistically satisfying a conservative taste for "romantic cliché, easy reference to past 'masterpieces,' and good décor" (272). These books may well be collecting dust on coffee tables as they serve these functions and testify to the good taste of their owners, but the poets who have been invited to write for them, many of whom work in or in close proximity to English departments,[19] are no doubt aware of this accusation. Many of their responses, I will show, reveal the dual ambivalence of wanting to accept the assignment while fending off charges of conservatism, and wanting to experience artworks aesthetically while addressing the problems of decontextualization, class privilege, and institutional control that Crimp summarizes.

All four of the sponsoring museums I address are academic institutions—two are part of major research universities (Yale and the University of Michigan), and the other two are national "academies" of art (the Art Institute of Chicago and the Tate). The forms of resistance we find in these anthologies reflect the assumptions of a late twentieth-century academic milieu in which critique is prized, and the differences among them reflect the differences in their institutional backings. *Transforming Vision,* coming out of one of the most prestigious museums in the U.S., presents more grandiose responses and less internal agitation than *A Visit to the Gallery,* which is more steeped in the terms of contemporary literary debate in the university. *Words for Images* maintains an urbane, intellectual tenor one would expect from Yale, and *With a Poet's Eye,* announced as a public enterprise, is more sensitive to the museum as emblem and purveyor of national values. If anthologies of ekphrastic poems were published by less mainstream museums, such as the Museo del Barrio in upper Manhattan, or the P.S. 1 Contemporary Art Center in Long Island City, poetic critiques of museums might look quite different, but book projects like these, with their high permissions and production costs, likely exceed the budgetary limits of smaller institutions. In this section, I assess the propensity for critique we find in mainstream museum-sponsored anthologies. Poets' attempts to confront the problems the museum setting raises, I argue, coexist uneasily with desires for and expectations about beauty and aesthetic response. This uneasiness often results in a poetic stalemate—a stalemate that manifests itself as vagueness or guardedness, formulaic reactions, skittish or glib conclusions, tonal flatness—when the critical impulse muffles acts of aesthetic attention.

Diane Wakoski's "Old Embroidered Chinese Robes in the Ann Arbor Museum" exposes crucial tensions in the museum experience but leaves their implications unresolved. The poem begins, confrontationally, by objecting to the class privilege on which a museum collection is founded:

> Sometimes I think that museums
> are just closets for the rich. (*Visit* 111)

The speaker is looking at articles of clothing, but in the museum they achieve the status of works of art, valued because they are beautiful luxury items once owned by members of an early twentieth-century Chinese imperial household. Choosing an object of the decorative or domestic arts, Wakoski shifts the ekphrastic enterprise into a category where social contexts are harder to ignore. She reminds us of the gap between the objects' museum presentation and their function in the life of the body: "Imagine / sleeping with the wearers / of these paintings? or watching them dress / in the morning?" She then undertakes the traditional ekphrastic task of presenting an object of captivating beauty:

> Stretched and pinned against the wall like a fan and not a
> garment
> is an old embroidered Chinese robe,
> as brilliant as this summer's urn
> filled with white and red
> impatiens, flowers
> waterfalling up and over its edges. (111)

Drawing comparisons with two similes ("like a fan," "as brilliant as this summer's urn"), Wakoski describes the robe's aesthetic effect. She thus begins the poem with two purposes, outlined in two free-verse stanzas: to expose the relation of the exhibited objects to class privilege, and to describe their aesthetic appeal.

Caught between these competing purposes, Wakoski suggests a connecting thread in the realm of the personal. Her speaker remembers that she owns a silk robe, stored in her "closet of baby / shoes and old fountain pens": "Like those other robes stretched against / the museum wall / my own robe, unworn, has become / an artifact / but there are no collectors to examine it [. . .]." Associating these personal examples of "unused beauty" and writerly self-reference ("old fountain pens") with those on public display, the speaker questions her social position as a writer, a position that authorizes her to go to museums and write poems about silk robes: "I ask myself what the difference is / between my closet and these museum walls / and know I will not find an answer [. . .]." Addressing the relation of the ekphrastic exchange to her own class status, she has nonetheless elided two significant differences: the garment in the museum is a "female theatrical robe," not a dressing gown, and it is on display for its rarity and beauty—because, for reasons inextricably linked to social stature, it commands greater aesthetic interest than her own robe. The poem does not investigate these aesthetic and functional differences, or the different social worlds they bespeak, and the indictment of the museum as a "closet for the rich" does not go beyond its initial observation: the museum experience is predicated on a set of privileged social conditions.

Molly Peacock's ekphrasis of Gilardi's *A Visit to the Gallery* (1877), the "title painting" of the Michigan anthology, also addresses the relation of the museum experience to social distinction, but the result is noncommittal. She imagines the thoughts of one of the lavishly dressed Victorian museumgoers in the painting, a young woman who longs to be rid of her stays and to escape the objectifying gaze of "a man's ideal." Attuned to the gendered dynamics of the scene, Peacock notes the class reality behind these women's activities: "may their servants stir / hot washtubs of bloody cotton strips to insure / they won't bleed on their taffetas" (*Visit* 27). The blithe tone and optative mood of this observation temper the critical impulse. Awareness of class distinctions does not unsettle the poem, and it proceeds along its formal course as a double sonnet. Peacock invites several questions about the relation of art appreciation to social conventions: What is the difference between the kinds of beauty implied by the "ruffled, laced, stockinged and corseted" Victorian women and the classical nude they admire? What does it mean to look at beauties looking at beauty, and how does the painting contrast beauty and ornament? But her highly structured poem notes the social underpinnings of the museum encounter only in a perfunctory way: recognition of class privilege does not extend to a fuller treatment of the underlying tensions that might destabilize the elaborate and reflexive artistry her own poem embodies.

In "Minimal Difference," Julie Ellison also attempts to account for the social and economic conditions that inform the museum experience, but she does so by forestalling the possibility of aesthetic pleasure. The tone of her poem, which addresses a scene engraved on an Inuit knife, is one of bureaucratic detachment. Ellison contextualizes her description of the object within the circumstances of a trip from "binocular" Canada to Michigan to visit the museum, a trip that is made possible through linguistic and economic forms of translation. Framing the ekphrasis in matters of taxation and transportation (the "GST," or Goods and Services Tax, the border), she finds an analog in the knife's scene of hunting and crossing, pointing out the ways the museum visitor goes through "customs" in two senses: the crossing of national borders requires a negotiation of touristic "differences," and the museum visit requires certain customary behaviors. The speaker comes from a "binocular" as well as bilingual perspective, and sees both the landscape of her travels and the scene on the knife as sites where the visual ("ice bar") and the auditory ("untranslated crunch") cross. This is as close to aesthetic perception as she is willing to tread—a glimpse of strange frozen motion where "jostled meringue" is perceived from a car window. For Ellison, the museum experience is part of a flux of disjointed traveling perceptions, and she stops short of engaging the object more fully by issuing this caution: "Hear your sighs of pleasure, class—/ pleasure like exhaust." Punning on "class" as schoolroom and social group, she critiques the "class trip" to the museum as exuding a suspect pleasure, a polluting taint (*Visit* 51).

A more compelling attempt to assimilate the attractions of the ekphrastic enterprise with the demands of museum critique is Laurence Goldstein's "Nydia, The Blind Flower Girl of Pompeii." Like Hayden, Goldstein positions himself on the museum threshold and observes that the museum keeps out grim world news, but he is much more ambivalent about this insularity:

> The big doors close out what's contemporary:
> the noise of a living culture, the earthshake
> of lurid bulletins, each a brutal crime.
> Always I glide first to this arresting figure,
> the wrought shape of Bulwer-Lytton's fancy,
> Nydia, flower-girl in the last days of Pompeii. (*Visit* 43)

As it was for Hayden, the museum visit is a habitual act of homage, but the work of art to which Goldstein returns is one that highlights the necessity of hearing, not ignoring, the bad news of the contemporary. Part and parcel of his homage is recognition of Nydia's lesson to heed disaster, a lesson given form by Randolph Rogers's 1858 sculpture of the main character from Sir Edward Bulwer-Lytton's 1852 novel. Through description of details of the sculpted form—Nydia's cupped ear and bent posture, the broken capital at her feet—Goldstein suggests that the work "signifies / the frailty of cities, their last extremity / when Nature throws down its ruinous fire." This work of art prompts reflection on the vulnerability of human civilizations and the disasters that befall them, not an escape from reality.

Nydia, as a "witness of world's end," compels Goldstein to consider the impact of artistic representations of apocalyptic events: what is the place, and efficacy, of acts of witness to human tragedy that we approach from within a museum? The question is one the anthology itself has quelled. When museum director William Hennessey welcomes the muses home in his preface to *A Visit to the Gallery* (Euterpe, Terpsichore, and Clio, muses of music, dancing, and history, have sponsored recent events), he adds that "Melpomene may be permanently excluded" (9). But Goldstein points out that the museum solicits a response to Tragedy even as it holds tragedy in abeyance:

> In museums catastrophe has a privileged place.
> All who walk through this temple of art
> suffer the need for some redemptive voice. (*Visit* 44)

In this poem, unlike Hayden's, the museum experience does not allow the poet to find a "redemptive voice." This "temple of art" spotlights and sanctifies "catastrophe," guiding its visitors through a ritual space where tragedy has been shaped formally and rhetorically, but the need for redemption continues to be suffered. Through an explication of the connotations of Rogers's sculpture, Goldstein explores the ways the museum neutralizes tragedy by suspending it at a historical distance.

Yet even as it raises these questions about the museum's relation to tragedies outside its walls, Goldstein's poem reveals an immediate disconnect between critique of the museum's insularity and the aesthetic experience the speaker seeks there. The forceful verb of the first sentence ("close out") becomes, in the second, a "glide" to an "arresting figure" of aesthetic captivation and form ("fancy," "wrought shape"). Aesthetic perception still provokes a kind of fade-out, a dissolving gaze: "this witness of world's end / pulls one, spellbound, close, closer . . . / / Twenty-five years I have visited you." The poem vacillates between these heightened moments of aesthetic affection, moments described as an I/you intimacy, and a more generalized voice of critique ("In museums catastrophe has a privileged place.") These shifts elide an unanswered question at the heart of the poem's treatment of "what's contemporary": what does a nineteenth-century literary-artistic representation of a historical tragedy like Pompeii have to do with more recent tragedies in which human beings, not nature, cause "ruinous fire"? The historical and aesthetic distance the museum imposes continues to complicate Goldstein's meditation. A quarter century of gazing at Nydia's "sealed orbs" leads him, in the poem's last stanza, "upward / where the imaginary smoke of extinction / gathers like the vanished clouds of 1945." The date "1945" drops a heavy weight into the poem, but the image is ambiguous. The "imaginary" smoke of a natural disaster, smoke that is *suggested* but not depicted by the work of art, evokes the "vanished clouds" of the Second World War. The comparison suggests that Goldstein is striving here to resolve ambivalence about the relation of art to historical violence. He tells us that Nydia "reminds the museumgoer how to look / at Death and Judgment waiting in the wings," but we are never sure how her reminder relates to the problem he poses in the opening and closing stanzas of the poem—that the museum and the aesthetic reverie it promotes inside its "big doors" mute the "earthshake" of contemporary events.

One of the most articulate museum critiques in these anthologies, one that clarifies many of the concerns that these poems raise, is Amy Clampitt's "The Song of the Lark" in *Transforming Vision*. Clampitt, however, chooses to present her reasoning in prose, and to quote one of her own poems as she challenges aesthetic mystifications. She does not reconcile critical and aesthetic

Figure 1. Jules Breton, *The Song of the Lark,* 1884.

demands, but further polarizes them as the domains of distinct genres, a separation that leaves critical awareness in one corner and aesthetic pleasure, impoverished by the rebuke, in the other. Addressing Jules Breton's *The Song of the Lark* (1884) (fig. 1), Clampitt implicitly responds to two other pieces about this painting that are included in *Transforming Vision*—excerpts from Willa Cather's

1895 essay "On Various Minor Painters," and from her 1915 novel *The Song of the Lark*. Read against each other, the texts by Clampitt and Cather forcefully illustrate the competing rhetorics that the anthology houses, in this case a direct confrontation of charisma with critique. Cather, whose character Thea goes to the Art Institute as a "place of retreat" and views Breton's painting with "boundless satisfaction" (*Transforming* 55), extols the painting's virtues in her essay: "You will find hundreds of merchants and farmer boys all over Nebraska and Kansas and Iowa who remember Jules Breton's beautiful 'Song of the Lark,' and perhaps the ugly little peasant girl standing barefoot among the wheat fields in the early morning has taught some of these people to hear the lark sing for themselves" (13). Clampitt, in her response, points out that "uplift of this specious sort is well known to have its own utility, with a dollar value in the offing" (58).

The attitude that Clampitt challenges here is one that frames the anthology to which she contributes. The excerpt from Cather's essay immediately follows the introduction, and it makes a grandiose claim on the museum's behalf (a claim that Hirsch quotes): "It is not unlikely that the Chicago Art Institute, with its splendid collection of casts and pictures, has done more for the people of the Middle West than any of the city's great industries" (*Transforming* 13). The museum, Cather explains, allows laborers to "find that these things of beauty are immortally joy-giving and immortally young" (13). These working-class visitors, in turn, offer "[s]ome of the most appreciative art criticisms I ever heard" (13). Her celebration of the museum expresses the belief that art is universally accessible and "immortally" powerful, as well as the belief that the uneducated museumgoer provides the purest interpretations.[20] When she goes on, somewhat contradictorily, to praise the museum for its role in propagating taste and culture among "the common people," urging well-to-do visitors from other cities to notice "the comparatively enlightened conversation of the people who frequent the building on free days" (13), she alludes to a contemporary controversy over museums' announced democratic missions and the reality of museum attendance. The Metropolitan Museum of Art in New York, for example, was criticized in the 1880s for its reluctance to open on Sundays, the one "free day" for the working class (Duncan 57–9). Writing a few years later, Cather boasts that "the spirit of caste is less perceptible in western cities": in Chicago, she claims, when a rich man exhibits a famous painting in the museum, "[his] workmen drop in to have a look at it some Sunday and decide that they would have done something better with the money, if it had been theirs" (13). Cather's attitudes here—her faith in artistic immortality and universality, her romanticization of working-class art perception—are a century old, but they nonetheless open and sanctify a collection of writings published in 1994.

The essay is accompanied by two photographs that underscore Cather's inau-
guratory role as the volume's presiding goddess-muse: a black and white
print from 1903/5 shows *Alma Mater* welcoming visitors to the grand lobby
of the Michigan Avenue entrance; beneath it, one of Rodin's Burghers of
Calais gazes on Adam in the galleries of European painting and sculpture
circa 1987.

A few pages later in the anthology, Clampitt revisits these galleries and
this ideological terrain. She remembers a framed print of *The Song of the
Lark* that her father, a hog farmer, brought home from the Art Institute in an
unusually profitable year. She then quotes from one of her own poems,
"Beethoven Opus 111," where a reference to this print, misattributing the
artist, appears:

> High art
> with a stiff neck—an upright Steinway
> bought in Chicago; a chromo of a Hobbema
> tree-avenue, of Millet's imagined peasant. . . . (*Transforming* 56)[21]

The next line of the poem, which Clampitt does not quote in the essay, notes
the painting's unseen titular subject: "the lark she listens to invisible, perhaps
/ irrelevant [. . .]." In this elegy for her father, the image of "Millet's imagined peasant" is one of several that juxtaposes his familiarity with hardship
with his appeals to "high art" for solace:

> diphtheria and scarlet fever
> every winter; drought, the Depression,
> a mortgage on the mortgage. High art
> as a susurrus, the silk and perfume
> of unsullied hands. (*Collected Poems* 51)

In her essay for *Transforming Vision*, Clampitt returns to the "imagined peasant" in this picture and to the problem her poem raises—those who have the
privilege to see laborers as lovely are those with "unsullied hands." Her critique begins in the form of self correction: the painting she remembers was
not by Millet, but Breton. She speculates that she erred because "Millet was
the bigger name," or because she remembered Edwin Markham's famous
ekphrasis of Millet's *The Man with the Hoe* (*Transforming* 56), and she
reopens the questions her earlier poem had occasioned.

From this perspective of self-critique, Clampitt addresses the painting
and romanticizations of labor in art. She offers this evocative prose ekphrasis

of Breton's painting with its ambiguous time of day: "[. . .] his barefoot peasant—head up, a sickle in her hand, and behind her, cloven by the horizon, an immense red sun" (*Transforming* 56). Though she agrees that the lark sings both morning and evening, she objects to interpretations of the painting as a laborer's late-day appreciation of nature: "I am disconcerted to be told in print that it is a *setting* sun" (56). She rejects the official ("in print") view, a view policed by what she calls "our aesthetic monitors":

> Surely, having stooped all day to wield that sickle, the limberest peasant would not be standing erect, or have bothered to listen for anything so far above the ground. If a setting sun is truly what the painter intended, my impulse is to say shame on him, because he must be lying—deliberately, and with the dubious intent of making up to somebody or other. (58)

Clampitt demands both realism and ethical accountability, objecting to what she sees as Breton's or his commentators' falsification of the circumstances of a field laborer's life. This kind of falsification in art, she admonishes, is obsequious and mercenary. In the beginning of her essay, she warns that she responds to the Art Institute's assignment for "all the extraneous reasons that are the despair of our aesthetic monitors" (56)—her ekphrasis involves extra-formal considerations, including social history, autobiography, and other forms of art. She uses her allotted space in the museum anthology to address the speciousness of attitudes represented by Breton's painting and its "official" aesthetic interpretations, and she openly resists the belief that art offers "visionary transformations."

But Clampitt's essay, with its stance of "shame on him," does not leave room for the point her poem has made about her father's difficult life: art has been sustenance nonetheless. The essay does not allow for the possibility that aesthetic experience has force and necessity beyond what the "aesthetic monitors" have dictated, that "aesthetic" might carry meaning other than 'that which willfully obfuscates or excludes the real.' Her poem does allow for these possibilities, and *not* through charismatic rapture. "Beethoven Opus 111" is deeply ambivalent about "hungers / for the levitations of the concert hall" (*Collected* 50), and about hungers for the elevations of the museum. It juxtaposes a man's appreciation for sensuous beauty—a flower's "luminousness / wounding the blank plains like desire" (52)—with accounts of a horrific case of poison ivy and his prolonged final illness. All of it is told in Clampitt's characteristically lavish language, a lyric density that makes its own running argument for sensory awareness and vivid detail. The poem concludes by holding out hope for art or music as a "levitation / of serenity," as "somehow reconstituting / the blister shirt of the intolerable / into these shakes and triplets, a hurrying / into flowering along the fencerows [. . .]" (52). I am perhaps just wishing Clampitt had sent Hirsch the poem instead, or at least that she had let her poetic voice speak longer as she sounded her objections. In her essay, agitation against the mystifying romanticizations represented by Cather's point of view leaves art and honesty at odds.

I read the weaknesses in the poems (and essay) I have discussed in this section as symptomatic of their incomplete assimilation of critical and aesthetic impulses in relation to the museum objects they survey. In their awkward junctures and hesitancies, they tend to isolate poetic imagery and extra-formal context, sensory description and discursive comment, as incompatible modes of thinking and writing. More successful critical interrogations in these anthologies, to which I now turn, reflect a greater willingness to allow these modes to interpenetrate, to reconcile competing demands. Varied in their approaches and styles, these poems have several characteristics in common: they are attuned to aesthetic response, they take a critical stance with regard to a problem raised by the museum setting, and they advance their critical positions through careful attention to artistic materials and their effects. Aesthetic experience, for these poets, is not immediate but *mediate:* it arises through perceptive engagements with the matter—the physical medium, the visual surface, and the critical issue—at hand.

MEDIATE RESPONSES: THREE CASE STUDIES

Alice Fulton's "Close," an ekphrasis of Joan Mitchell's *White Territory* (1970–1) in the University of Michigan Museum, matches Mitchell's turbulent visual surface with an edgy verbal one.[22] The painting, which critics describe as a gestural evocation of the winter landscape of the French town of Vétheuil, where Monet lived and Mitchell settled in the 1970s (*Visit* 125n, Dixon 2), draws Fulton into the orbit of "second generation" abstract expressionism and its concerns—spontaneity, dynamic execution, monumentality and mystery, an improvisatory and urgent engagement with artistic materials. Fulton's response to the painting is emotive and engrossed in this urgency, but it is also an uneasy reaction to the museum atmosphere that frames her experience of it, an atmosphere that abstract expressionism, with its requirement of expansive and immaculate spaces for aesthetic contemplation, helped create: "the standards it set—of scale, intensity, and inwardness—still determine much modern art, and by extension, the liminal ambience of permanent museum collections" (Duncan 130–1). By foregrounding the museum setting of her encounter with Mitchell's painting, and emphasizing that her aesthetic response is a function of an engagement with its material, Fulton interrogates three "charismatic" standards for art and its exhibition—art as larger-than-life (its scale), art as an intuitive means to immediate emotion (its intensity), and art as occupying a liminal, isolate sphere (its inwardness).

Fulton begins by pointing out that the painting's scale (it measures seven by nine feet) complicates the museum's attempt to exhibit it:

> To take it further would mean dismantling doorframes,
> so they unpacked the painting's cool chromatics
> where it stood, shrouded in gray tarpaulin
> near a stairwell in a space so tight
> I couldn't get away from it.
> I could see only parts of the whole
> I was so close.
>
> I was almost in the painting,
> a yin-driven, frost-driven thing
> of mineral tints
> in the museum's vinegar light.
> To get any distance, the canvas or I
> would have to fall down the stairs
> or dissolve through a wall.
> (*Visit* 71)[23]

Because of its size, the museum staff must leave the painting in a space adjacent to service facilities ("near a stairwell"), an ironic compromise of the "aesthetic hang" that abstract art traditionally calls for: "The wish for ever closer encounters with art [has] gradually made galleries more intimate, increased the amount of empty wall space between works, brought works nearer to eye level, and caused each work to be lit individually" (Duncan 17). "Close" takes this wish for "ever closer encounters with art" to its ridiculous extreme. This painting is isolated, but not in an airy aesthetic chapel. The "intimate" response it elicits derives not from its autonomy, but from its uncomfortable proximity to other structures and to the body of the observer herself. The effect is akin to that of Vito Acconci's *Proximity Piece* (1970), in which Acconci stood uncomfortably close to museumgoers looking at exhibits. The experience Fulton describes likewise "disturbs the museum visitor's expectation of a sort of contemplative privacy" (McShine 21). The word "so" marks her discomfort twice: the space is "so tight" and the painting "so close" that the typical museumgoing procedure—stepping back and getting a good look at the painting as a whole—is not possible.

Instead, the speaker's position—"almost in the painting"—provokes a vertiginous interaction with its surface effects and physical environment. Unaccustomed to such proximity, the speaker has difficulty getting her bearings, a difficulty emphasized by the right justification, which shifts us off balance from the outset, and by the four successive lines beginning with "I." A fusillade of "in" sounds (yin, frost-driven, mineral, vinegar) underscores the speaker's sense of immersion "in" the details of the painting's surface. The phrases "shrouded in gray tarpaulin" and "the museum's vinegar light" suggest that the museum drapes those details in a mausoleal pall and suspends them in a preservative, even as the "cool chromatics" cast their own light. With the word "shrouded," Fulton gestures toward a central point of museum critique, the objection to its funereal atmosphere and anesthetizing effect on perception. In Adorno's formulation of this objection, the museum embalms its contents as it elevates them: "The German word, '*museal*' has unpleasant overtones. It describes objects to which the observer no longer has a vital relationship and which are in the process of dying" (175). Fulton's speaker, caught in an atmosphere that is at once deadening and disorienting, revitalizes her approach to the artwork by noticing the internal and external factors that color it. The encounter with the painting is not simply one of a museumgoer beholding an art object, but one in which both beholder and painting quite literally bump into the museum: "To get any distance, the canvas or I / would have to fall down the stairs / or dissolve through a wall" (*Visit* 71). The line-break on "I" emphasizes the speaker's precarious position: the physical and institutional environment of the museum (walls and stairs, installation and illumination) constrains the exchange it enables, and the beholder must continually renegotiate her position with respect to that environment.

Both of the melodramatic possibilities Fulton proposes here—falling or dissolving—remind us that the notion that art suddenly knocks us off our feet, or sparks some mystical communion, is incompatible with the actual conditions of the museum encounter. Her speaker does not experience the painting through an immediate and intuitive descent of grace, but through a strenuous, dialectical process of describing the artist's medium and technique. Mitchell's canvas, with its squared patches of periwinkle, pale green, and olive-black, and its central knot of russet wisps, suggests both a muted winter landscape and what Mitchell called "internal weather" (Dixon 2):

It put me in mind of winter,

a yin-driven enigma and thought
made frost. When I doused the fluorescents
it only became brighter.
The background spoke up
in bitter lungs of bruise and eucharist.
Of subspectrum—
(*Visit* 71)

The speaker explains that the painting "put me in mind of winter," a phrase that preserves an ordinary idiom for 'it made me think of winter' while invoking Wallace Stevens's directions in "The Snow Man" for radical numbness in the face of austerity. As a mysterious "yin-driven" embodiment of "thought / made frost," the painting appears to test the limits of such numbness. Its cold brightness is barely tolerable, and the museumgoer reaches for the light switch to "douse" the institutional fluorescents, only to discover that the painting "became brighter," its glare coming from within. Seeing the painting is thus a process of observation and allusion, of modulation and adjustment of both perspective and environment.

This process generates the odd image of "bitter lungs of bruise and eucharist," terms that are linked more by acoustic affinity than sense. As soon as she drifts toward the suggestion of sacramental significance, however, Fulton hesitates, revising the "of" clause that modifies "bitter lungs" to replace "eucharist" with "subspectrum," a term from science, not ritual. The poem continues to broach and resist the possibility of aesthetic liminality— of art's isolate and potentially epiphanic inwardness. The speaker "almost" participates in a lyrical idealization of the painting's intrinsic clarity and authenticity:

I was almost in its reticence

of night window and dry ice, its meadow
lyric barbed in gold, almost
in the gem residence
where oils bristle into facets
seen only in the original, invisible in
the plate or slide [. . .].
(*Visit* 71)

The images "night window," "dry ice," and "meadow / lyric barbed in gold" interpret the painting as an atmospheric Romantic landscape. Mitchell herself suggested this interpretation when she said that a "territory" (like this "White Territory") "is a lyric space" (Dixon 2). Fulton emphasizes this lyricism when she describes the painting's "reticence" and "gem residence"—a perfectly faceted realm apart. But this interpretation becomes strained when the speaker notices, in abruptly end-stopped lines, that the painting

> [. . .] shrinks to *winsome* in a book.
> Its surface flattens to sleek.
> In person, it looked a little dirty.
>   (*Visit* 72)

In a reversal of the usual understanding of the difference between original and copy, she notes that the original is dingy, whereas the reproduction is "sleek." The reproduction, Fulton cautions anthology readers, has more of an aura of polished art than the original.[24] Arriving at these discoveries, revising "gem residence" with "a little dirty," the ekphrasis becomes a record of ambivalence about lyricism and originality.

Fulton ultimately stresses the value of seeing the original painting not because it is more perfectly authentic, but because its texture records the artist's encounter with her material:

> I could see the artist's hairs
> in the pigment—traces of her
> head or dog or brush.
> [. . . .]
> I saw how turpentine had lifted the skin,
> leaving a ring, how the wet was kept
> on the trajectories, the gooey gobs of
> process painted in. Saw dripping
>
> made fixed and nerves and
> varicosities visible.
> I saw she used a bit of knife
> and left some gesso showing through [. . .]
>   (*Visit* 72)

The "close" encounter the speaker has with the painting enables her to see that art is adulterated by evidence of the processes that have made it. The artist's body or brush (or even her dog) has left behind particles that mark the surface, a surface that itself has a "bodily" character with its "skin," "nerves," and "varicosities." (Later in the poem, we get an image of "winter tissue and cranial-colored paint.") Evidence of the interaction of paint and turpentine allows the speaker to see "process painted in," and the unpainted parts of the canvas reveal that this artistic project is not one of coverage or closure, but restless contingency.

The speaker's uncomfortable awareness of the "closeness" of this museum encounter leads to awareness of the closeness of the composition process itself:

> While painting, she could get no farther away
> than arm's length.
> (*Visit* 72)

The speaker sees that the distance of an "arm's length"—far enough away to see, close enough to touch the canvas—is the restriction that enables artistic production. Recognition of this difficult but enabling position allows the speaker to acknowledge the painting's power and its making—to embrace an abstract expressionist painting while rejecting the requirement of liminal ambience. In "the museum's vinegar light," the speaker perceives the painting not in the quietude of its isolation, but in the discomfort of its proximity to herself and to the museum frame. Its emotional force derives not from its immediate effect on the speaker, but from a series of jostling approaches. The speaker appreciates the painting as a work of art not through sudden insight, but through a highly self-conscious effort to understand the processes and materials that created it.

The museum frame around a work of abstract art invites similar wariness in Charles Wright's "Summer Storm." Like Fulton, Wright examines an issue central to the modernist museum and the aesthetic approach it promotes—the concern that the museum freezes and moderates the meanings of the works it exhibits by severing their connections to the world outside the world of art. Wright considers this problem by testing the extreme case, a painting that seems to be purified of external reference and to achieve a purely formal significance, Piet Mondrian's *Composition—Gray Red* (1935) in the Art Institute of Chicago. If he can find referents in Mondrian, then the museum's walls are more porous, more susceptible to the extremities of history, than an ideal of abstraction would allow. Wright's poem addresses tensions between representation and abstraction, fullness and vacancy, history and stasis: his ekphrasis is a critique of formalism through an inquiry into the nature of formalism, a critique that is negotiated through attention to aesthetic response and artistic medium.

Most critics have emphasized Wright's visionary and mystical preoccupations, but here his subject is the problematic relation of art to historical and political violence.[25] In this poem, which like the painting exudes a sense of high-voltage control, the noise of war erupts to dispel museal reverence, and to ground visionary ambitions. The poem begins with a summary of Mondrian's aesthetic stance:

> As Mondrian knew,
> Art is the image of an image of an image,
> More vacant, more transparent
> With each repeat and slough:
> > one skin, two skins, it comes clear [. . .] (*Transforming* 115)[26]

Wright glosses Mondrian's process as one of distillation and "sloughing" of factual connotations to achieve greater clarity, but his poem works in the opposite direction, accumulating facts as it shifts from generalization to ekphrastic description:

> Two rectangles, red and gray, from 1935,
> Distant thunder like distant thunder—
> Howitzer shells, large
> > drop-offs into drumbeat and roll. (*Transforming* 115)

Wright offers a succinct formal analysis, stating the number and color of geometric shapes in the painting. To these visual details he adds a date, a place on a historical timeline: with this addition, the objective particulars lead to other nouns that occupy, not vacate, the painting's space as a field of significance, images that take on a life that the painting's perpendicular frames cannot contain.

The mention of the date of the painting's composition sparks an emotional response and an extra-formal investigation. The poem enters the realm of comparison, suggesting that the painting is metaphorically evocative of an atmosphere of "distant thunder," and also that the speaker has turned away from his objective analysis to remark on an external stimulus, the "summer storm" of the poem's title. The comma after "1935" allows for both readings, but the metaphor is quickly reduced to tautology, the possibility of resemblance yielding only an identity equation in the form x = x. With a dash, however, Wright ratchets up the tension, introducing "Howitzer shells" at the exact midpoint of the poem. The line-break and indentation underscore the jarring "drop-offs," and the onomatopoeic values of "drumbeat and roll" provide sonic emphasis for this shift into martial territory. With this series of spare gestures, the poem opens itself, and the painting, to historical and political interpretations, however oblique. The German army used Howitzers to batter Belgium in 1914, when Mondrian was prevented by war from returning to Paris. By 1936, when this painting was completed, the German threat that would provoke Mondrian to leave Paris in 1938 was already ominous. Without making any direct reference, Wright invites us to view the painting's painful precision and austerity as an abstracted and re-abstracted reflection of Mondrian's response to war in Europe.

Violent upheaval is suggested most strongly by the painting's single colored field, a red rectangle at the edge of the canvas in the lower right. Six times longer than it is wide, the rectangle suggests that we are seeing the edge of a large red square that remains almost out of view. Wright focuses our attention on this red space. Despite his characterization of Mondrian's aesthetic as one of "vacancy" and tautological abstraction, Wright cannot help but seek referents, and the ones he offers are as far from aesthetic neutrality as possible:

> And there's that maple again,
> Head like an African Ice Age queen, full-leafed and lipped.
>
> Behind her, like clear weather,
> Mondrian's window gives out
>   onto ontology, [. . .] (*Transforming* 115)

We find ourselves in a turmoil of interpretation. What maple? How can we be seeing it "again"? Then, with a provocative and racialized image, Wright rejects Mondrian's rejection of external subject matter. Instead, we find the red rectangle transformed into an autumnal-maple-tree-cum-African-queen, lavishly described in alliterative adjectives ("full-leafed and lipped"). "Mondrian's window" appears "behind her," suggesting that the particularity of the poem's proper names—Mondrian, Howitzer, African—trumps the process of rarifying abstraction. The reductive transparency of "the image of an image of an image" has become a window that "gives out / onto ontology"— an abstract category indeed, representing all that exists as well as the inquiry into all that exists, but even this vastly generalizing term does not dim the graphic vividness of the images that precede it.

In this brief (18-line) poem, Wright demonstrates that Mondrian's "window" of art "gives out" in a triple sense: it succumbs to a fatigue of representation, distilling the world into aesthetic principles; it frames and portrays a world beyond it nonetheless; it bestows and disseminates its meanings. We may not have any idea what exactly the painting has to do with an African Ice Age queen, but naming this image has irrevocably altered our perception of the red rectangle. Wright concludes by drawing attention to the painter's medium:

> A dab of red, a dab of gray, white interstices.
> You can't see the same thing twice,
> As Mondrian knew. (*Transforming* 115)

Wright leaves us, not with a notion of abstraction or of purified aesthetic transparency, but with the opacity of Mondrian's material—"dabs" that mark the presence of paint, and "interstices" that mark its absence. The poem is formally symmetrical, the final line repeating the first to underscore that repetition effects both resemblance and transformation. We are brought back from the thunder of Howitzer shells at the poem's center to the artistic enterprise announced at its inception, and are reminded that the speaker is standing before a painting, contemplating its surface as well as its resonance. Maintaining a tense equilibrium, Wright's poem dramatizes a tension between an aspiration for rarefied autonomy, and the pressure of historical and political meanings that reverberate through the museum's walls.

Both Fulton's and Wright's poems address artworks from the beholder's singular viewpoint, but Stanley Kunitz's "The Sea, That Has No Ending," after Philip Guston's *Green Sea* (1976) in the Art Institute of Chicago, presents charisma and critique in dramatic dialogue from the perspective of the painting. Adapting the traditional ekphrastic strategy suggested by Simonides's statement that "a poem is a speaking picture,"[27] Kunitz ventriloquizes the strange figures in Guston's painting. In the process, he queries notions of aesthetic mystery and the myth of the uncreated creator, and challenges the interpretive authority of the museum.[28] He takes as epigraph a descriptive note from *Master Paintings in The Art Institute of Chicago*:

> *Green Sea* is one of a series of paintings [Philip] Guston did in 1976 fea-
> turing a tangle of disembodied legs, bent at the knees and wearing flat,
> ungainly shoes, grouped on the horizon of a deep green sea against a
> salmon-colored backdrop. . . . Its meaning eludes us. (*Transforming* 98,
> Kunitz's ellipsis)

Juxtaposing the official text of the institution against his own poem,
Kunitz steps in where the museum concedes interpretive defeat, but his
ekphrasis does not attempt to resolve the painting's elusive significance.
Instead, he offers a wry commentary on the idea of authority as it arises
within and around this "Master Painting." In the introduction to *Transforming Vision*, Hirsch mentions Kunitz's poem and cites its concluding
lines to pre-empt the criticism that poems in the collection are "merely"
aesthetic: "there is something large and fundamental at stake in most of
these pieces. They attend closely to artistic considerations [. . .] but they
are not aestheticized. 'This is not an exhibition,' the artist storms at the
conclusion of Stanley Kunitz's metaphysical lyric 'The Sea, That Has No
Ending,' 'it's a life!'" (11). Kunitz's poem, as we will see, makes a more
complicated point about the nature of aestheticization and exhibition
than Hirsch allows.

Giving voice to this "scruffy tribe," a group of legs and feet "huddled
on this desolate shore, / so curiously chopped and joined," Kunitz observes
that the life these figures bespeak cannot be extricated from the painting's
enigmatic iconography. The speakers reflect on the losses their disembodied
limbs suggest:

> How many years have passed
> since we owned keys to a door,
> had friends, walked down familiar streets
> and answered to a name? We try
> not to remember the places
> where we left pieces of ourselves
> along the way, whether in ditches
> at the sides of foreign roads
> or under signs that spell "For Hire"
> or naked between the sheets in cheap
> motels. Does anybody care?
> All the villagers have fled
> from the sorry sight of us. (*Transforming* 98)[29]

In the interrogative, the speakers recall markers of lost lives—homes, friendships, known geographies—but even signifiers as fundamental as names have been stripped away, leaving no ties to a citizenry or knowable reality outside the painting. Reminiscence is quickly curtailed by an effort "not to remember" the places where these ties were severed: three prepositional phrases ("in ditches," "under signs," "between the sheets") suggest forms of political, economic, and sexual exploitation, but as soon as these losses are sketched, the speakers reject the notion that they can be communicated in any meaningful way—"Does anybody care?" Disconnected from the "villagers" who "have fled," they are only a "sorry sight," a painted image on a museum wall.

The speakers then reject the possibility that their suffering is authored or authorized by a Creator:

> Once we had faith that the Master,
> whose invisible presence fills the air
> watching us day and night, would hear
> our cries and prove compassionate,
> but we are baffled by his words
> even more than by his silences. (*Transforming* 98)

The painting occasions a shrugging version of a Job-like complaint to an inscrutable God, a lament that is not addressed to the Master, but rather to us, listening in from the vantage of beholding the painting. The "Master" behind this "Master Painting in the Art Institute of Chicago," an echo that the proximity of the epigraph makes us hear, suggests not only God but the Artist—the one who imagined these figures and put them in their dismal array on the shore of "the sea, that has no ending." Kunitz presents the painting as an artistic parable from which no clear principle can be extrapolated, a vision of the sour defiance of figures whose "meaning," determined by a Master-God-Artist, is not known to themselves or to anyone who attempts to understand what they represent. In response to this skeptical renunciation, Kunitz introduces a second voice, italicizing the declarations of the Master raging against his creations:

> But he who reads our secret thoughts
> rebukes us, saying: *You cannot hope*
> *to be restored unless you dare*
> *to plunge head-down into the mystery*
> *and there confront the beasts*
> *that prowl on the ocean floor.* (98)

Kunitz sets up a dialogue in which skeptical voices confront a grandiose rhetoric of restoration, risk, and mystery. The speakers immediately undercut the Master's declaration by suggesting with scare quotes that his naming is arbitrary: "'Sacred monsters' is what he calls them." With their deadpan Beckett-esque resignation, they reject any possibility of salvific "grand heroic action." They will take no Keatsian plunge into mystery, nor exert themselves in a struggle on behalf of the "sacred." By giving voice to this defiance, Kunitz suggests that the painting foils attempts to ascribe visionary or symbolic significance to it. The speakers ask, "Why is the Master knocking at our ears, / demanding immediate attention?" and their refusal to answer to creative authority is also a refusal to be interpreted. The painting frustrates any hope for immediate apprehension of a message, or motive, behind its livid tones and crude shapes.

Staging this mini-rebellion against the charismatic ideology, Kunitz allegorizes the museum experience: the Master whose "invisible presence fills the air / watching us day and night" also suggests the atmosphere and surveillance of the museum itself. Near the end of the poem, the speakers remark colloquially,

> It's really strange how much we miss
> those people who came to gape and jeer;
> we'd welcome their return, for company. (*Transforming* 98)

Ironically, the speakers miss the "villagers" they have turned away, longing for some interpretive community in which they might communicate their predicament. The phrase "those people" suggests uncomprehending museumgoers who happen upon the painting, respond with astonishment or ridicule, and abandon it. The "we" of the poem remains an assemblage of cartoon joints and thickly cobbled soles, propped on a green slab. Kunitz starts to tell "their" story only to document a narratival dead end—there's no one left to hear the tale. The Master then returns in a demonic guise, indignant that the speakers should take this nonchalant attitude toward their obsolescence:

> In the acid of his voice we sense
> the horns swelling at his temples
> and little drops of spittle
> bubbling at the corners of his mouth.
> *This is not an exhibition,* he storms,
> *it's a life!* (98)

With the word "exhibition," Kunitz ends on a reflexive turn, foregrounding the museum setting and taking us back to the epigraph, with its concession that the painting exhibits only its elusiveness.

Hirsch, in his introduction, quotes these final lines as evidence that a poem about art is not necessarily limited to aesthetic concerns, that "life" impinges on aesthetic experience. But the lines he cites are spoken in the furious voice of the Master with "little drops of spittle / bubbling at the corners of his mouth." They are not a justification of art's relevance to life, but the impotent ravings of a Master who has lost control of his recalcitrant creation. (These lines are also a playful jab at an artist Kunitz knew well, and whom he once described as prone to "excited talk, with little pockets of moisture bubbling at the corners of his mouth" [qtd. in Balken 65].) Concluding his poem with these lines, Kunitz makes the point that this painting *is* an exhibition, not a life at all. He draws our attention to the very fact of this work's "aestheticization"—its bewildering opacity, arbitrary signage, and representational red herrings. By framing his ekphrasis in a museum scenario and the museum's language of commentary, he presents an attitude that the Master finds scandalous—the "exhibition" does not reveal a life, but only an exhibition. Hirsch, anxious to ward off the charge of aesthetic insularity, misses Kunitz's critique: insistence on art as a transparent medium through which we can see "life" presupposes the concept of a stable, authorizing imagination (a "Master"), and a concept of a stable, coherent interpretation (recognition as a "Master Painting"), two concepts that Guston's painting, with its disjunctures and disturbances, undermines. *Green Sea* is an exasperating picture, and a compelling one, Kunitz suggests, because it resists transcendent significance and relies stubbornly on its remoteness—it occupies a space apart but not a liminal space, a space where rapture and mystery give way to an aesthetic of mordant tenacity.

For these three poets, the occasion of an ekphrastic poem opens a reflexive space for interrogating the mediations—the media, processes, and contexts—that enable, structure, and complicate aesthetic experience. Fulton's proximate scrutiny of Mitchell's materials, Wright's echoes and expansions of Mondrian, and Kunitz's peculiar colloquy all represent poetic attempts to cross a museum threshold into aesthetic intensity while refusing the trance of idolatry and rapture—they are attempts to see with both passion and clarity. Without sacrificing the fullness and particularity of their attractions to and insights into works of art, these poets respond to the problems of the museum setting: they confront, from different angles, the ideal of aesthetic autonomy, the museum as cultural mortuary, the separation of art objects from their social and political contexts, the effects of institutional control, the apotheoses of originality and formal perfection, the concepts of master artist and masterpiece, and the authority of interpretation. These poems, and these anthologies, are valuable for their internal friction, their tolerance of ambivalence and incongruity, and their multiplication of aesthetic and critical possibilities in response to works of art in museums.

For many readers, however, these critiques will not have gone far enough. Many will feel that these poets' critical challenges do not outweigh their complicity with the ideals and expectations of the institution that sponsors them. These poems ignore, for the most part, the class structure intrinsic to high-cultural distinction and leave its hierarchical values intact. They do not question the instrumentality of language in articulating personal experiences, or the stability of the poetic speaking voice. They are (merely or nonetheless, depending on your point of view) subtle, reflective, perceptive poems about works of art by prize-winning poets whose projects are individual but not, in the usual terms of contemporary poetry criticism, "innovative"—they have not radically altered the traditional parameters of the ekphrastic enterprise formally, syntactically, or ideologically. In the next chapter, I examine the ekphrastic work of a poet who writes under the sign of innovation, but whose meditations in museum space expose the tensions that inhere in that very term.

Chapter Two

"Returned Again to the Exhibition": John Ashbery, Avant-Gardism, and Ekphrastic Risk

Dozens of poems in Ashbery's oeuvre reflect encounters with visual media. Throughout a career that spans 50 years and 23 volumes of poetry, Ashbery has borrowed artists' imagery and lexicon, taken titles from art generically ("The Painter," "The New Realism," "Statuary") and specifically (de Chirico's *Double Dream of Spring*, David's *The Tennis Court Oath*), mentioned artists (Caspar David Friedrich, Overbeck, Millet, Caravaggio), and collaborated with them (Joe Brainard, Jonathan Lasker, Elizabeth Murray). His critics have long recognized the importance of this influence, and explored the diverse instances in which the visual arts inform his "logic / Of strange position" (MSO 56)[1]—his characteristic poetic mode of transcribing the obliquities, undertones, and intervals of perception as he goes.[2] But critics have overlooked the ways Ashbery's "strange position" vis-à-vis the visual arts makes him particularly responsive to the distractions and framings of the museum setting. As "Tapestry" explains,

> It is difficult to separate the tapestry
> From the room or loom which takes precedence over it.
> For it must always be frontal and yet to one side. (AWK 90)

The site of exhibition and the site of creation—the "room" where the tapestry is displayed, and the "loom" where it is made—demand notice. They "precede" and even take priority over the work, superimposing a perceptual movement from frontal gaze to glimpse askance. This peripheral vision is a difficulty, a complication of the impulse to separate the *objet d'art* from its surroundings and its making, but for Ashbery it is a necessary complication. As he writes elsewhere, this context of room and loom, institution and artistic process, conditions any viewing of an aesthetic object: "True, it is only a picture, but someone framed and hung it; / it is apposite" (HL 82). Curatorial intentions, matters of framing and hanging, give a picture its contiguity with its surroundings and also its aptness, its pertinence. Ashbery's approaches to pictures treat these surroundings in apposition to the works they frame, invoking a "logic of strange position" that reflects both aesthetic response and critical appraisal.

Most discussions of Ashbery's career emphasize not his peripheral vision but his double vision—his seemingly binocular contribution to American poetry. Criticism of his work has been governed, and hampered, by the idea that there are "two John Ashberys" (Lolordo 750), a dual characterization employed to explain how his status as the best poet of his age—"the most universally acknowledged of poets writing in English" (Lolordo 755), "the most influential poet writing in English" (Herd 1), "the most widely acclaimed living American poet" (Kellogg 103)—can be announced from divergent positions in the contemporary poetic field.[3] According to this double profile, "Mainstream Ashbery" behaves in accordance with the Romantic-to-High-Modernist tradition, and is "best read in a temporal line of succession," such as the one offered by Harold Bloom (Emerson, Whitman, Stevens) or Helen Vendler (Wordsworth, Keats, Tennyson, Stevens, Eliot) (Lolordo 752). This Ashbery is credited with writing *Self-Portrait in a Convex Mirror* (1975), and most of *Double Dream of Spring* (1970). His other more unruly personality is promoted from the "aleatory base camp" (I borrow Ashbery's own terms [CY 49]), where critics sympathetic to Language writing trace his "degree of disjunctiveness" (McHale 564) as a measure of his experimentalism. Finding it waxing and waning from book to book, they devise narratives of progress and relapse, stressing that "Avant-Garde Ashbery" (who was included in Allen's *The New American Poetry*) wrote his second book, *The Tennis Court Oath* (1962), went underground, and later re-emerged with renewed vanguard energy in volumes such as *Flow Chart* (1991) and *Hotel Lautréamont* (1992).[4] Ashbery's contribution and his prominence, this account of dual canonization suggests, are a function of his ability to work at different times within mutually exclusive poetic paradigms.[5]

In this chapter, I argue that except as critical exaggerations, there are not and never were two John Ashberys.[6] The one John Ashbery who should command our attention, because he complicates and interrogates the very terms of these critical differences, is Ashbery the museumgoer. When we find him in the museum, in his poetry and in his more than 30 years of art criticism,[7] we find him engaged in ambivalent rethinking of the binary parameters of mainstream traditions and avant-garde traditions, of ideas of continuity, pastness, conservation, closure, and homage on the one hand, and ideas of rupture, presentness, innovation, openness, and experiment on the other. In moods that range from nostalgic to vituperative, he challenges the terms of his own reception as they are mirrored and magnified in this particular cultural site and the artworks he finds there. The equivocation and doubt that surface in museum scenes spanning Ashbery's career, I claim, reflect a fundamental ambivalence about the opposition between tradition and avant-garde itself, an ambivalence that is for Ashbery initiating and sustaining rather than compensatory. To make this case, I begin by considering Ashbery's problematic relation to the ekphrastic tradition, and then show how two museum scenes in his recent work engage a tension between the interrogative present of ekphrastic looking and the weight of cultural belatedness. Then, in the second section, I explain how attention to Ashbery as a museumgoer enables us to recalibrate our understanding of his relation to avant-gardism. I reexamine the ways innovation was inseparable from its institutionalization in the cultural moment in which New York School poetry began, and then analyze a museum scene in Ashbery's early work that reflects the uncertainties of this juncture. Finally, I argue that his most famous ekphrastic poem, "Self-Portrait in a Convex Mirror," is far more ambivalent about "tradition" than critics have assumed. Considered in the light of—or rather, in the glare of—the 1984 Arion edition, in which the poem appears in a stainless steel canister with original prints by contemporary artists, "Self-Portrait in a Convex Mirror" raises complex questions about the attitudes of homage and resistance that its museum setting exposes.

"IN THE NIGHT OF THE MUSEUM": ASHBERY AND THE EKPHRASTIC TRADITION

Writing an ekphrastic poem carries a potential risk for an avant-garde poet: it invokes a relation to past art and prior representation that could be seen as running counter to a commitment to experiment. It is retrospective, an act of looking at something already made, as the conventional epigraph indicates—"after Caravaggio." Ekphrasis often appeals to the traditionalist because it has a long literary history, from Homer to Keats to Auden, and because it pays homage to an act of aesthetic creation that has endured. It can be rejected by the experimentalist on the same grounds: you can't be *avant* if you are *après*. In a recent article, Brian McHale points out that ekphrasis is one of Ashbery's dominant modes, observing that many descriptions in his poetry, "the ontological rug having been pulled out from under them, are retrospectively reframed as ekphrases" (562). Seemingly straightforward descriptions turn out to be representations of visual artifacts—paintings, maps, even jigsaw puzzles. But in making the case that ekphrastic "secondhandedness" exemplifies Ashbery's late-postmodern poetics of "'recycled' language, [. . .] appropriated, mediated, simulacral materials" (563), McHale overlooks the problems that ekphrasis as a traditional genre of "retrospective reframing" poses for Ashbery. A poetic process of "secondhand" appropriation, especially when the materials appropriated are other works of art, can also carry a "traditional" valence, and this orientation toward the revered past provokes deep ambivalence in Ashbery's ekphrastic work. He often gives us ekphrasis at one further "ontological remove" than McHale acknowledges: when he situates the ekphrastic encounter in the museum, he draws a frame around the frame, exposing the tensions between visual priority and poetic immediacy that inhere in his approaches to visual art.

Ashbery directly addresses this "ekphrastic risk" in the midcareer poem "And *Ut Pictura Poesis* is Her Name." As he explained in his introductory remarks at a reading of this poem in 1989, the phrase *ut pictura poesis,* from Horace's *Ars Poetica,* indicates that "a poem should be like a painting" (qtd. in Lisk 36). The poem quarrels with this directive from the first, its title enclosing the bookish term in a flippant contemporary idiom, and the first line insisting "You can't say it that way anymore" (HD 45). Declaring it out of date, Ashbery simultaneously invokes and rejects "the tradition in which Horace's phrase *ut pictura poesis* serves as an injunction to guide a poetic art that, in spite of its resistant verbal resources, seeks to emulate the spatial and visual arts [. . .]" (Krieger 78). Commanding "You can't say it that way anymore" challenges this traditional paradigm of emulation in which the visual arts come first and the poet follows. The pattern is difficult to escape. Even the evocative figure for the generative potential of writing that Ashbery offers in this poem—"The extreme austerity of an almost empty mind / Colliding with the lush, Rousseau-like foliage of its desire to communicate" (HD 45)—follows this sequence: the adjectival use of the proper name "Rousseau" points back to a prior artistic effort with a particular visual signature and aesthetic appeal. But the poem continues to resist this sequence, offering advice "About what to put in your poem-painting" (HD 45). The term "poem-painting," the hyphen yoking the nouns into a single syntactic unit, conjoins the two media to suggest simultaneity rather than emulation. Giving instructions for creating this hybrid form, Ashbery puts the two media on the level:

Flowers are always nice, particularly delphinium.
Names of boys you once knew and their sleds,
Skyrockets are good—do they still exist?
There are a lot of other things of the same quality
As those I've mentioned. Now one must
Find a few important words, and a lot of low-keyed,
Dull-sounding ones. (HD 45)

He recommends the inclusion of visual data (delphinium, skyrockets, sleds), but he also partially severs the tie to prior visual referents by emphasizing that the *names* of boys are to be included, and then a selection of *words*. In this characteristic gesture of foregrounding the verbal medium itself, Ashbery suggests an ekphrastic mode that sidesteps the trajectory of visual first and verbal second. He stresses that "as in painting" does not necessarily mean 'as in a pre-existing painting' or even 'as in a pre-existing visual scene,' but 'in the same way that a painting might be made,' by accretion, juxtaposition, and balance.

Reading *ut pictura poesis* this way, Ashbery is reverting to a meaning of Horace's text that precedes its adoption as the banner of the pictorialist tradition in poetry. Murray Krieger reminds us that this tradition is based on a misreading: "in the passage in question Horace is doing no more than considering similarities in audience (or reader) response to works from different arts: some [. . .] to be seen close up and some from a distance [. . .]" (79). Horace is making the point that the overall quality of a work changes the way he perceives local felicities or flaws:

sic mihi qui multum cessat fit Choerilus ille,
quem bis terve bonum cum risu miror; et idem
indignor quandoque bonus dormitat Homerus;
verum operi longo fas est obrepere somnum.
ut pictura poesis: erit quae si propius stes
te capiat magis, et quaedam si longius abstes. (39)

[so I find that a writer who is frequently at fault becomes for me a Choerilus whom I gape and giggle at if he is good two or three times; yet equally I am put out when Homer, who is regularly good, nods; but in a long work one must admit the incursion of sleep. Poetry is like painting: one painting will captivate you more if you stand closer and another if you stand further off.] (58)[8]

Jean Hagstrum explains that scholars' punctuation of the sentence so that a comma *followed* the verb "erit" ("it will be that")—presenting the phrase as "ut pictura poesis erit"—skewed the statement in such a way as to strengthen its meaning as a maxim, not simply an observation: "When read with the first clause, the verb makes Horace's meaning seem more dogmatic than it actually is: 'a poem *will be* like a painting'" (60). Ashbery does not address this translation controversy directly, but he seems to recommend a reading of Horace's text that is truer to its occasional context—*ut pictura poesis* as comparison of particular examples, as cognitive exchange and verbal-visual mediation.

Nonetheless, Ashbery is not willing to renounce the *ut pictura poesis* tradition qua tradition out of hand, even if it means a retrospective motion that might have concerns about belatedness in tow. The counter-stance in the history of inter-art relations—the insistence on the generic and practical separation of verbal and visual arts—holds little appeal for a poet so fruitfully provoked by painting. It is tempting to read a passing reference in a late poem to a "sunset tie" given to the speaker by "old Mrs. Lessing" (SWS 72) as a joking allusion to this tradition. The widow of the most adamant antagonist of the sister-arts tradition, one surmises, gives the poet a kitsch landscape with the advice to wear it, not write it. (It is not entirely absurd to speculate that Ashbery had Lessing's insistence on the distinctive spheres of the arts in mind here. The Laocoön group comes up later in the same volume, and the poem in which Mrs. Lessing appears flouts the separation of visual and verbal arts by bringing together Mutt and Jeff comics and Hölderlin.) Ashbery's basic objection to Lessing's division of the arts is clear, as he states in an essay on Saul Steinberg, after questioning why painting should be considered non-narratival and poetry non-sensory: "eyes, minds and feelings do not exist in isolated compartments but are part of each other, constantly crosscutting, consulting and reinforcing each other" (RS 280). Ashbery adopts a position midway between *ut pictura poesis* and the *paragone,* between the sister arts and the arts in combat and isolation. His speculations on the influence of painting on poetry and vice versa repeatedly supplant notions of competition or dominance with mediating terms, as in an essay on R. B. Kitaj: "How wonderful it would be if a painter could unite the inexhaustibility of poetry with the concreteness of painting" (RS 307). By stressing reciprocal exchange, Ashbery avoids the "which came first" predicament. Instead, he celebrates poetry's "inexhaustibility" without the usual possessive maneuver of claiming this capacity as language's domain: he offers it, albeit conditionally, to Kitaj.

Ashbery's own strategy for utilizing ekphrasis without consigning it to a retrospective engagement with the art of the past draws on this same notion of "inexhaustibility." He transforms a liability into an asset by employing ekphrasis as a prolific mode of description, interpretation, apostrophe, response, and counter-response—an always approximate and ongoing rather than precise or complete treatment of its object. He does not attempt to produce a verbal icon that can contain tensions within a formal unity—the high-modernist aspiration for the mode. Instead, he turns to ekphrasis as a way to repel closure and coherence, a way to go on and on. In so doing, he makes ekphrasis available to the avant-garde effort to produce and reproduce an ongoing present. For Ashbery, ekphrasis entails a process of finding multiple textual correlatives, comments, and queries for pictorial artifacts, a process always on the verge of getting out of hand, as in this passage from "Valentine":

> Seen through an oval frame, one of the walls of a parlor. The wallpaper is a conventionalized pattern, the sliced okra and star-anise one, held together with crudely gummed links of different colored paper, among which purple predominates, stamped over a flocked background of grisaille shepherdesses and dogs urinating against fire hydrants. To reflect on the consummate skill with which the artist has rendered the drops as they bounce off the hydrant and collect in a gleaming sun-yellow pool below the curb is a sobering experience. Only the shelf of the mantelpiece shows. At each end, seated on pedestals turned slightly away from one another, two aristocratic bisque figures, a boy in delicate cerise and a girl in cornflower blue. Their shadows join in a grotesque silhouette. In the center, an ancient clock whose tick acts as the metronome for the sound of their high voices. Presently the mouths of the figures open and shut, after the mode of ordinary conversation. (HD 64)

This humorous ekphrasis of an illustrated valentine card draws on the "inexhaustibility of poetry" indeed. An abundance of color words (five) and prepositional phrases to indicate spatial relations (13) suggest that the verbal could be employed to map and comment on the visual ad infinitum. The ekphrasis ends only with the seemingly arbitrary notice of the figures' mouths, which remind the speaker of the verbal realm of "ordinary conversation" and lead him into examples of such. Ashbery occasionally offers a minimalist counter-gesture against this kind of ekphrastic excess, as in a mini-paragone, one of "37 Haiku," where an approach to visual art is quickly curtailed by subtractive textual work: "[. . .] original artworks hanging on the walls oh I said edit" (W 37). But by and large, Ashbery is much more prone to the opposite tendency—ekphrastic inexhaustibility as counter to belatedness.

This strategy has not pleased all readers. Its consequence is prolixity, nowhere more apparent than in Ashbery's *Girls on the Run* (1999), which at 55 pages is possibly the longest single ekphrastic poem ever written. The poem reflects the scale to which Ashbery has tended in recent work. He has published nine volumes of new poetry since 1991, six of these running to over a hundred pages. This copious output has fatigued most of his commentators, leading the least sympathetic of them, like William Logan, to conclude that Ashbery has "succumbed to a continual and unstoppable collapse" (327). Others find redeeming moments in the onslaught, as when Marjorie Perloff cites *Can You Hear, Bird* (1995) as evidence that success has not spoiled him ("Normalizing" 1). Part of many critics' impatience with the later Ashbery, I believe, is that under either of the reigning regimes in contemporary poetry, the story of his career has already been written: one story consolidates his legacy as heir of high modernism, and the other measures his modes of experimentation against notions of exemplary postmodernism. To unsettle both of these stories, I turn to the mostly uncharted territory of Ashbery's late work, where ekphrastic undertakings in two equivocal museum scenes expose the difficult intersection of traditional and innovative artistic agendas.[9]

In *Girls on the Run*, Ashbery takes the ekphrastic risk of writing *after*, but only by addressing a work that cannot be considered traditional *or* avant-garde, and by spinning a vast web of description, narration, and interpretation around its images. Ashbery's subject is the work of Henry Darger, an "outsider" artist whose illustrations for his 15,000-page saga of the warfare and espionage of the "Vivian Girls" have earned him posthumous acclaim as the second great modern primitive (after Henri Rousseau).[10] Darger's extra-canonical position—he is neither mainstream artist nor iconoclast per se—propels Ashbery's experiment, an exhaustive ekphrasis that suggests narrative movement without conclusion, symbolic shimmer without seriousness. Charmed by the antics of Judy, Pete, Dimples and the rest, he offers passages such as this one:

> Pink shrouds fell on the pansy jamboree,
> mocking the circular nature of events with its own kind of back-to-the-
>   beginning
> free fall. A few pansies got drowned. Yet this was as nothing to the terrible
> muttering of the distant cavalry [. . .]. (GR 43)

Copiousness and superfluity are at the core of this enterprise: why write a 55-page poem to address artworks that are already illustrations of a 15,000-page novel? Admirers will say "why not?" and moderately tolerant readers, myself included, will enjoy the verve of the first dozen pages, only to regret that the poem softens into familiar Ashbery demi-lyrics and muddled discourses on all things dreamed or overheard. What makes this attempt significant and revealing, I would argue, is that it represents one of the few *openly* ekphrastic poems in Ashbery's work (others are "Self-Portrait in a Convex Mirror" and "Double Dream of Spring"), and the *only* time, to my knowledge, that he employs the conventional epigraphic "after," telling us after his title, flush right, that he writes "after Henry Darger."

Taking this risk, the poem complicates its own belatedness in a scene of self-reflexive and inventive museumgoing. In the midst of the ekphrastic profusion, Ashbery pauses and draws full attention to the museum setting in which the inspiration takes place:

> Nov. 7. Returned again to the exhibition. How strange it is that when we least imagine we are enjoying themselves, a shaft of reason will bedazzle us. Then it's up to us, or at any rate them, to think ourselves out of the muddle and in so doing turn up whole again on the shore, impeached by a sigh, so that the whole balcony of spectators goes whizzing past, out of control, on a collision course with destiny and the bridesmaids' sobbing. Of course, we listened, then whistled, and nobody answered, at least it seemed nobody did. The silence was so intense there might have been a sound moving around in it, but we knew nothing of that. Then we came to. The pictures are so nice on the walls, it seems one might destroy something by even looking at them; the tendency is to ignore by walking around the partition into a small, cramped space that is flooded with daylight. (GR 12–13)

A repeated visit to a museum or gallery leads the speaker to observe that the work's appeal has been unexpected but epiphanic, almost hallucinatory. The pronoun slippage in "we are enjoying themselves" conflates object-directed perception (enjoying them, the artworks) and self-directed recreation (enjoying ourselves), suggesting that the museum visit invites not only attention to the paintings' details and effects, but also a blurring of the distinction between art and audience. The speaker is personal, writerly, taking up the inspiring moment of "bedazzlement" as a responsibility "to think ourselves out of the muddle" and launch his own imaginative flights. The moment is a call to cognitive and creative restoration, a longing to "turn up whole again," but it is a call that cannot be disentangled from the pronominal confusion of "us" and "them." Ashbery draws attention to the ways the institutional setting fosters concentrated scrutiny of the paintings and also prompts an imperative for further creative work.

The phrase "impeached by a sigh," suggesting an accusation by exasperation, initiates a cascade of non sequiturs. The moment of bedazzlement in the exhibition—of aesthetic apprehension and admiration—leads not to turning up "whole again" but to a bizarre high-speed fiction of museumgoers "whizzing past, out of control." The clichéd "collision course with destiny" spins into the soap-operatic sobbing bridesmaids, then gives way to an isolation chamber—the museum evacuated of all sound except the solitary whistle. To keep moving past this moment of silence, Ashbery plays the "it's all a dream" trick: "Then we came to." After the swoon of rapture, the acceleration of melodrama, and the eeriness of silence, the speaker regains consciousness and resumes normal museum decorum: "The pictures are so nice on the walls." Noticing the lighting and the mode of display, he turns away from the ekphrastic project as if to get his bearings in the present of his own weird perceptions, framing the ekphrasis in the "impeaching" recognition of the mental and institutional circumstances that interrupt it. The poem goes back to Bunny and Philip and their tale, but this museum-conscious passage introduces a pocket of uncertainty and indeterminacy. *Girls on the Run* is no ordinary ekphrasis, Ashbery insists here, but a dramatic engagement with both an existent work and a set of mercurial responses to it, an engagement that must remain fluid, unexpected, and imaginatively resourceful.

In his next book, *Your Name Here* (2000), Ashbery presents another equivocal museum scene that unsettles homage to prior visual art with the tenuousness and uncertainty of artistic discovery. "Caravaggio and His Followers" opens with a museumgoer's perfunctory remarks:

> You are my most favorite artist. Though I know
> very little about your work. Some of your followers I know:
> Mattia Preti, who toiled so hard to so little
> effect (though it was enough). Luca Giordano, involved
> with some of the darkest reds ever painted, and lucent greens,
> thought he had discovered the secret of the foxgloves.
> But it was too late. They had already disappeared
> because they had been planted in some other place. (YN 19)

The poem deftly negotiates several changes in tone, from the clichéd enthusiasm of "most favorite" and "nice," to the eloquent richness of "lucent greens" and "foxgloves," to melancholic reflection at the end of the poem. The speaker begins with effusive appreciation for "my most favorite artist," but he turns immediately to Caravaggio's successors—Preti, whose production is both paltry and sufficient, and Giordano, whose superlative vividness broaches a "secret" insight but proves to be belated and elusive. In halting syntax that suggests the museumgoer's starts and stops, the poem begins by paying homage to a line of artistic descent, but quickly steps into the next gallery and forgets the precedent. The speaker is more interested in what artistic developments will happen next, such as Giordano's strenuous "involvement" with his medium, which Ashbery underscores by linking "Luca Giordano" visually and sonically with "lucent greens." Giordano, who was nicknamed "Proteus" for his skillful pastiches of other artists' styles, holds more attraction for this museumgoer than the dramatic chiaroscuro of Caravaggio, the alleged favorite, even if Giordano's efforts lead only to failure, to discovery being pre-empted. Giordano occupies a more obscure place in the gallery, a stop between the illustrious predecessor and the innovations of "some other place."

The poem continues with a meandering and disjunctive ekphrasis. In characteristic Ashbery fashion, the second stanza suggests the accumulation of a museumgoer's cognitive scraps, glancing notice of a painting or paintings, and overheard conversation, enacting the process "of how we go along, first interested by one thing and then another [. . .]" (YN 19). Indecisive observations about color shift from art to life and back: in one painting, tree bark is "the color of a roan, / perhaps, or an Irish setter"; elsewhere, a streetlight "changes to green," then to a painted garment's "sapphire folds." European painting is filtered through particularly American perceptions: "I say, if you were toting hay up the side of a stack / of it, that might be Italian. Or then again, not. / We have these things in Iowa, / too [. . .]" (YN 19). Counterbalancing the grandiose museum presentation of the "old, old secret of the foxgloves" are moments of American folk wisdom: "But, as Henny Penny said to Turkey Lurkey, something / is hovering over us [. . .]" (YN 19–20). In the midst of these ambling perceptions and interpretations, the passage focuses into evocative ekphrastic glimpses, as in the image of "a shotsilk bodice Luca Giordano might have bothered with" (YN 19). Ashbery sharpens this image with the repetition of short "o" sounds and the echo of "bodice" and "bothered"—the line itself is shot through with a sensory pattern that is both precise and colloquial.

The final stanza of the poem, however, undermines these ekphrastic fragments and ironizes the earlier homage by foregrounding the museum setting. In an inversion of the usual terms of ekphrastic looking, the point of view shifts from the museumgoer to the paintings themselves, in the gallery after-hours:

In the night of the museum, though, some whisper like stars
when the guards have gone home, talking freely to one another.
"Why did that man stare, and stare? All afternoon it seemed he stared
at me, though he obviously saw nothing. Only a fragment of a vision
of a lost love, next to a pool. I couldn't deal with it
much longer, but luckily I didn't have to. The experience
is ending. The time for standing to one side is near
now, very near." (YN 20)

Ashbery stages a fantasy of unchaperoned access to the museum's contents, as
if the artworks "talk freely" once the guards and visitors have left. A careful
chain of partial rhymes harmonizes these voices—stars, guards, stare, experience, near—even as the long lines and simple diction suggest casual banter.
Literalizing the Simonidian principle that "a poem is a speaking picture," but
without employing any description that we can be sure refers to Caravaggio
or his followers at all, Ashbery gives these paintings a voice of irritable
remonstrance. What these speaking pictures say is that we have gotten them
all wrong, deluded perhaps by some narcissistic nostalgia. Even the most per-
severant attempt to see penetratingly—to "stare, and stare"—results only in
blindness: "he obviously saw nothing." In this ventriloquial aside, the paintings speak up to correct the mistake of one who would romanticize them in a
moment of museum rapture. The colloquial phrase "I couldn't deal with it"
suggests a rolling of the eyes at the maudlin interpretation of "a fragment of a
vision / of a lost love, next to a pool." Rather than ekphrastic illumination,
we get this strangely opaque backtalk.

   Asking "Why did that man stare, and stare?" these quoted speakers
challenge a stance that Ashbery has relied on throughout his writing life, in
poetry and in prose. He has arguably stared in museums more than any of
his contemporaries (Frank O'Hara might have surpassed him, but Ashbery
has had 39 more years). "Why did that man stare, and stare?" is a quiet self-
admonishment. As this poem moves from the thoughts of a naive museumgoer, to a perceptual collage, to these rebuffing personae, we have the
unnerving sense that the "man" who was doing this unendurably romantic
gazing is also the speaker imagining this twilight museum scene, framing it
in a kind of Caravaggesque tenebrism where "some whisper like stars." Setting the scene "in the night of the museum," rather than the more ordinary
phrasing "at night in the museum," suggests an elegiac overture: more is at
stake than the museum's closing time. The poem concludes with a fatalistic
sense that "the experience is ending," the present perfect tense and the phrase
"near, very near" suggesting that the "end"—of art? of the speaker's life?—is

asymptotically closer if not yet upon us. Here, "in the night of the museum," the mode of "standing to one side" and seeing askance that Ashbery has cultivated throughout his career—the mode of "Tapestry"—suggests not only seeing peripherally or ironically, but also an imminent ceding of position, a stepping down.

In light of this ending, the poem's coalescence of matters of canonical precedent and innovation, enraptured appreciation and obsolescence, points toward Ashbery's reflections on his own canonization: now that he has obtained "most favorite artist" status in his own medium and era, has his "logic of strange position" become one more relic of the cultural past? The title of this volume, *Your Name Here,* suggests both a fixed place in the poetic record and complete anonymity, a phrase from "personalizable" goods. Another poem in the book offers this humorous lament: "He's gone, who never lacked for champions, / killed by daylight saving time, or a terrible syllabus accident" (YN 37). These pre-emptively posthumous concerns are not new in Ashbery's work, but they appear with increasing frequency in the later books. In *Girls on the Run,* we hear: "Your talents are warehoused now" (GR 46). In the book before that, his poetic gifts have become an arthritic symptom: "my negative capability acting up" (Wk 14). Eight books back, in 1992, a speaker wants to figure out the nature of his vocation and "staunch the energy hemorrhaging" from his career (HL 26). Ashbery's later work is preoccupied with dread and faux-dread of a "terrible syllabus accident"— would such an accident forget him, or put him in the wrong company, or skew the story of his discoveries? As he writes in *Flow Chart* (1991): "We [. . .] have only sterile notions of staying included to ruffle through, and one never tires of this retrograde motion, even as one fears the consequences of standing still and becoming like an old chromo on a wall" (FC 56). "Caravaggio and His Followers" gives voice to those old pictures on the wall, allowing them, and perhaps even canonical poets, a chance to keep backtalking their admirers. The doubting and accusations of misreading that we find here reflect Ashbery's ambivalence about the opposition between "retrograde motion" and its opposite, between "standing still" and moving forward artistically. He expresses this ambivalence by pausing in the museum, the institutional setting where innovation becomes tradition, where "discovery" is codified in a form that becomes, for posterity, the work of someone's "most favorite artist."

These are the tactics of a museum-going Scheherazade. Ashbery utilizes a museum-conscious ekphrastic mode to transform belatedness into an interrogative present, sustaining the acts of attention, description, and interpretation that ekphrasis requires as a function of their inexhaustibility. It is a strategy that is, whether we appreciate the long-winded results or not, a productive one—a source of unusual angles and juxtapositions, fresh perceptions, new poems. I have begun this chapter with these two late instances because they present a tension that I see as fundamental in Ashbery's work. Ten books back, right after the publication of his *Selected Poems* (1995), this tension arises in another museal atmosphere, in a poem titled "Never to Get It Really Right": "A tan light stalks the rooms now / (With their neoclassic moldings, waiting / For the tedium of words to subside), / That suits it all by draining the life out of it [. . .]" (AG 68). Again the gallery, the very architecture, waits for the wordmakers to go home, and a predatory light embalms the rooms' contents. The tone is elegiac:

> Besides, the gold of winter is clanging already
> In dusty hallways. I have my notebook ready.
> And the richly falling light will transform us
> Then, into mute and privileged spectators. (AG 68)

Though a museal paralysis threatens to set in, the light transforming "us" into "mute and privileged spectators," the speaker is ready to keep writing. The "tedium of words" refuses to subside. As the poem proceeds, the spectators become "just bemused revelers, no / Qualms, no frustrations, not even a sense of its being new" (AG 69). The poem presents one of many such gestures in Ashbery's work—the has-been poet bidding farewell, the avant-garde poet kneeling in the institution of canonization—as if continually invoking and ironizing such gestures staves off the need for them. They are gestures to keep in mind when we examine his early career's "sense of its being new."

THE ELABORATE PAST AND THE OUTERMOST BRINK: THE MUSEUM AND AVANT-GARDE AMBITION

In a poem from *Hotel Lautréamont* (1992) entitled "Of Dreams and Dreaming," Ashbery expresses his doubts about "the idea of museums":

> Meanwhile we live in the paperweight of swirling blizzards
> and little toy buses painted vermilion like the sky
> when it rises up reasonably to our defense in the half-hour
> after sunrise or before sunset and likes to, it likes
> the idea of museums. Then so much of us is fetched away.
> Often you think you can see or even smell some part of it
> before it too is put away, used and put away. But then these

so recent nights would be part of the elaborate past, that old
contraption, the one we were never sure about— (HL 90)

This poem presents Ashbery as miniaturist, a mode that Dan Chiasson points
out he "learned from Elizabeth Bishop, and perhaps from Joseph Cornell
[. . .]" (144). It is a mode that is comfortable in museums, yet as the scene
unfolds, from snow-globe to protective sky, fondness for museums provokes a
peculiar stutter: the syntax halts after the "to" that would announce an infini-
tive, perhaps an active desire to "go" to the museum, and doubles back to
indicate instead that "it likes / the *idea* of museums." This moment of hesita-
tion and abstraction sparks a concern about being "used and put away," about
the present ("recent nights") becoming part of "the elaborate past." The concern is given tentative resolution in the poem's final line: "Besides, when in
doubt you can strike a match" (HL 92). This concluding image perfectly captures an ambivalent stance toward "the elaborate past"—a match can be
struck to illuminate the darkness, to allow for some discernment, or it can be
struck to burn it all down. "Of Dreams and Dreaming" reminds us of Ashbery's continued suspicion of "the elaborate past, that old / contraption, the
one we were never sure about," a suspicion that the "idea of museums" is
incompatible with the idea of being "avant-garde."

    Discussions of avant-gardism in recent years have repeatedly stressed that
an equivocal stance inheres in any attempt to be avant-garde after the "historical
avant-garde," the movements in the 1920s associated with vanguard politics in
Berlin and Moscow, including Dada, surrealism, futurism, and constructivism.[11]
Yet discussions of Ashbery's early experimental work and his avant-gardism more
generally have tended to take at face value the myth of origins he offers in an
essay entitled "The Invisible Avant-Garde." In this essay, an address to the Yale
Art School that appeared in *ARTnews Annual* in 1968, Ashbery criticizes the
contemporary artistic climate where it is "safest to experiment" and the "avant-
garde has absorbed most of the army" (RS 393, 392). He addresses a dilemma
that has since become so familiar that it is considered an axiom of the postmodern period: as Rasula observes, "The perpetual quandary of the avant-garde is its
dialectical relation with the conventional because, through time, the renunciations of the vanguard become new conventions. The 'postmodern' [. . .] begins
by acknowledging this dilemma, which has the status within the postmodern of
one of Zeno's paradoxes" (245–6). But even as Ashbery addresses this paradox in
this speech, he takes a decidedly un-postmodern tack—he posits a moment in
his own lifetime when it did not apply. In this section, I argue that rather than
reverting to the narrative of prelapsarian avant-garde origins that Ashbery offers
in this instance, we should recognize that ambivalence about avant-gardism and

its institutionalization informs his work almost from the beginning, an ambivalence that is, moreover, directly proportional to his proximity to museums.

Facing the "big booming avant-garde juggernaut" of the sixties, Ashbery asks how an artist can sustain an effort to experiment and innovate, and whether public approval negates that effort: "Is there nothing then between the extremes of Levittown and Haight-Ashbury, between an avant-garde which has become a tradition and a tradition which is no longer one?" (RS 393) The speech suggests the importance of finding a stance that avoids either orthodoxy, but it also includes this grouse:

> Things were very different twenty years ago when I [. . .] was begin-
> ning to experiment with poetry. At that time it was the art and literature
> of the Establishment that were traditional. There was in fact almost no
> experimental poetry being written in this country, unless you counted
> the rather pale attempts of a handful of poets who were trying to imitate
> some of the effects of the French Surrealists. The situation was a little
> different in the other arts. Painters like Jackson Pollock had not yet been
> discovered by the mass magazines—this was to come a little later,
> though in fact *Life* did in 1949 print an article on Pollock, showing
> some of his large drip paintings and satirically asking whether he was
> the greatest living painter in America. [. . .] I found the avant-garde
> very exciting, just as the young do today, but the difference was that in
> 1950 there was no sure proof of the existence of the avant-garde. To
> experiment was to have the feeling that one was poised on some outer-
> most brink. In other words if one wanted to depart, even moderately,
> from the norm, one was taking one's life—one's life as an artist—into
> one's hands. (RS 390)

The capital "E" on establishment, the hyperbolic claim that there was "almost no experimental poetry," and the melodramatic appeal to a kind of metaphysical absence—"no sure proof of the existence of the avant-garde"— enhance the romantic claim that his youthful vocation entailed intrepid pioneering, a life or death struggle.

This scene of "invisible" opposition has become something of a sacred text for critics eager to establish and celebrate Ashbery's avant-garde pedigree. David Sweet, for example, states that in this essay Ashbery

> offered his most comprehensive statement regarding his relationship to
> the Avant-Garde by tracing the outlines of an 'invisible avant-garde' that

presumably included him. For Ashbery, the true Avant-Garde only properly exists in a condition of cultural tenuousness: unconsolidated and largely unrecognized, disestablished less out of a social hostility against it than uncertainty about it, its quality always that of *yet becoming*. ("Ut Pictura" 319–20)

Establishing this "cultural tenuousness" is crucial to critical accounts that stress the gulf between the mainstream tradition and the avant-garde in postwar American poetry. An attempt to underscore the "invisibility" of Ashbery's early avant-garde work occurs in Marjorie Perloff's 1997 essay "Normalizing John Ashbery," where she claims that the mainstream discovered him only belatedly: "Indeed, Ashbery attained almost no recognition prior to the publication of *Self-Portrait in a Convex Mirror*, published in 1976 when the poet was fifty. It was only after the relatively accessible title poem of this volume became well-known, that the Establishment started to come around" (5). There is much evidence to the contrary. W. H. Auden selected Ashbery's first book, *Some Trees*, for the Yale Younger Poets Series in 1956. Ashbery's second book, *The Tennis Court Oath*, the key text for critics who place him in a language-centered experimental genealogy, was selected by John Hollander for the Wesleyan Poetry Series in 1962. By the time *Self-Portrait in a Convex Mirror* appeared, Ashbery had already published five books and appeared in 19 anthologies, including *The Norton Anthology of Modern Poetry* (1973) (Rasula 493).[12]

Sweet's account of Ashbery's avant-gardism in relation to the visual arts is more nuanced and balanced, but in the end it falls into the same idealization of the avant-garde that Perloff's does. Observing that Ashbery often draws on both the "regressive" and the "progressive" in the techniques of historical vanguard movements, and "find[s] avant-garde potentialities even within the most traditional means" ("Ut Pictura" 316, 319), Sweet observes that the opposition between tradition and avant-garde is unstable in Ashbery's poetics:

> [T]he 'other tradition' to which he so often refers [. . .] [may] designate his own eccentric middle way, a 'new' negative capability that skews and ultimately reverses the meaning of the terms 'tradition' and 'avant-garde.' Ashbery assumes a nomadic tangency in relation to these two discursive poles of poetic production—one, established/decaying, the other, emergent/proliferant. As these poles increasingly come to resemble each other in a culture that seems to celebrate, commercialize, and consume even the most radical forms of art, Ashbery's quiet, mannered

idealism becomes a private holdout against the forces of cultural homogenization. (320–1)

The phrase "radical forms of art" marks the point where Sweet finds himself at a loss, tangled in the two poles he shows "to resemble each other." Even as he claims that Ashbery's work functions at a tangent to both poles, he clings to the idea that a "true Avant-Garde [. . .] properly exists" (320). Praising Ashbery's "quiet, mannered idealism" as a "private holdout" against commercialism is an empty gesture—Ashbery hardly escapes cultural enshrinement by being "mannered," as we will see clearly in the next section. Retreating from his observation that the two poles intertwine, Sweet goes on to maintain the myth of the pure Avant-Garde: "The key virtue of a true avant-garde poet, then, is disinterestedness rather than antagonism" (321). Later he remarks, "the only true avant-garde stance is that in which the last thing said is always already outdated" (322). As the word "true" in these passages indicates, even in this final instance with its aporetic finesse, the idea of an authentic avant-garde at the heart of Ashbery's practice holds powerful sway over his critics.

"The Invisible Avant-Garde" is cited more frequently than any of Ashbery's art writings, but the equivocation of this text has been glossed over: even as he declares it "invisible," Ashbery hedges on the matter of the canonization, indeed the musealization, of an innovative artistic project (Kenneth Koch will confront this matter directly, as I discuss in the next chapter). Owing in part to the rhetorical occasion of the speech, an address to art students who probably considered the forties ancient history, Ashbery is eager to contend that the New York School painters, for whom his coterie of poets is named, were "authentically" avant-garde at the time he was inspired by them. But in the midst of a clutter of padding phrases ("a little different," "though in fact," "still a long way," "a little later"), Ashbery mentions the 1949 *Life* article on Pollock. He is asserting that the promotion of Pollock to the mainstream had not yet happened, so he declares the article "satirical" to make his point. But a look at the *Life* article tells another story. Although the tone is clearly bemused, and *Life*'s writers maintain only pseudo-objectivity in noting (notoriously) that "others" consider Pollock's work "yesterday's macaroni" and "an aerial view of Siberia," there is nonetheless an unmistakable endorsement at the end of the piece: "Even so, Pollock, at the age of 37, has burst forth as the shining new phenomenon of American art. Pollock was virtually unknown in 1944. Now his paintings hang in five U.S. museums and 40 private collections" (qtd. in Altshuler 165). Ashbery's complaint about the co-option and normalization of innovation belies his own insistence: he cites an article that records and *quantifies* the entrance of avant-garde art into museums.

To clarify this inaugural moment in Ashbery's career and the anxieties it raises, I want to stress that the coming together of the New York School poets as a group dedicated to innovation and experimentation corresponds not to an "invisible avant-garde" but to the period of institutionalization of the avant-garde for which they are named.[13] As early as 1946, while Ashbery was still an underclassman at Harvard, the marketability of action painting was already being remarked upon: Mark Rothko wrote to Barnett Newman that "Pollock is a self contained and sustained advertising concern" (qtd. in Clark 304). Cecil Beaton's famous photographs of fashion models in front of Pollock's paintings appeared in *Vogue* in 1951 (Clark 304), the year of the Ninth Street Show, which is often considered the height of New York School painting's downtown, bohemian, vanguard spirit (Altshuler 159). Ashbery and his friends, now living in New York, no doubt caught this fever. But the same year Ad Reinhardt was already mocking "Look, Ma! no hands" gestural technique as "Museum Racing Form." The race for the museum was on. Mrs. John D. Rockefeller III was derided in *Time* in 1951 for buying abstract paintings, but she was buying them, and *Harper's Bazaar*, *The New Yorker*, and *Vogue* were presenting favorable views. Widespread commercial success began to take place by mid decade: "de Kooning's 1956 sold out exhibition of landscape-like abstractions at Janis [is] often mentioned as initiating the new age" (167). By the end of the decade, the downtown scene's "primary concern was whether the advances seen on 9th Street had calcified into a new academy" (173). In 1959, *Life* did a two-part feature on "abstract expressionists, world's dominant artists today" (173).

Even though Ashbery had left New York for Paris in 1955—the point is commonly made that he did not witness the establishment of this new academy first hand, as did Frank O'Hara, Kenneth Koch, James Schuyler, and Barbara Guest—he began to work as an art critic shortly after his arrival in Europe and was hardly out of touch. On the contrary, his time abroad reflects a *conscious* decision to get away from the all-too-obvious institutionalization of the New York avant-garde. Part of the appeal of expatriate existence, as he writes in "American Sanctuary in Paris," is precisely the *"feeling"* that one can escape from the institutional machinery of canonization, from the "split-second of [a work's] trajectory from easel to gallery to museum" (RS 88). Ashbery could not help being aware that he and his friends were linked to the "world's dominant artists today." By the time John Bernard Myers, director of the Tibor de Nagy Gallery, invented the name "New York School of Poets" in 1961, "hoping to cash in on the cachet of the world-conquering Abstract Expressionists" (Lehman, *Last* 20), the name "was a half-joke, hedged from the start by discomfort and irony" (G. Ward 277). As is particularly clear in hindsight, abstract expressionism no longer represented a vanguard position as soon as it was "funded by a US State Department avid for promotion of a new American art that would express the nation's Cold War supremacy in cultural terms" (G. Ward 278). O'Hara, though he undoubtedly did not perceive these artists to be tools of Cold War hegemony, participated directly in this process as curator at MoMA, organizing a traveling exhibition of the work of Pollock and others (Perloff, *O'Hara* 91). Ashbery saw this show in Paris in 1959 (RS 103).

Much later, when the need to insist on how hard it was to be avant-garde in the old days was perhaps less pressing, Ashbery himself made the point I have underscored. In 1986, in an essay on Jane Freilicher, he admits: "I hadn't realized it, but my arrival in New York coincided with the cresting of the 'heroic' period of Abstract Expressionism" (RS 240). Even as early as 1964 he refers to "Pollock & Co.," glibly acknowledging that this particular avant-garde had been quickly "incorporated" into the commercial mainstream.[14] His later recollections stress the importance of *museumgoing,* rather than renegade opposition, to his early poetic experimentation. He writes that major shows of works by Munch, Soutine, Vuillard, and Matisse at MoMA in the late forties and early fifties affected his artist friends like "catnip," and inspired him as well: "I began pushing my poems around and standing words on end" (RS 241). MoMA in the forties was still the home of the outrageous and the new (as it still is for many today), but Ashbery's memories of these "historic" "big shows" remind us that this museum's halls were already hallowed. (All of these artists, with the exception of Matisse, had died in the early or mid forties). Remarking on his youthful need for museum "stimulation," Ashbery remembers that what had already become artistic *tradition*— the canonical account of turn-of-the-century and early-twentieth-century modern masters in the "permanent collection" at the MoMA—compelled a frenzy of experimentation. The "idea of museums" might have been a reminder of cultural consecration and artistic energy "used and put away," but museums still provoked a practical and immediate effort to realize the possibilities of his medium.

The avant-garde then, for Ashbery, is always already institutionalized, and the primary site of this institutionalization, the museum, was a backdrop of his work from the beginning. In "The Skaters" (1963), a 30-page poem written at the moment in his career when he became not only a habitual but a professional museumgoer, a museum setting frames an odd contest between avant-garde ambition and the "elaborate past." It is a poem about (among other things) poetic identity formation and the posture one adopts with respect to the past and future. The excerpts from it that Ashbery includes in his *Selected Poems* (1985) give us a familiar portrait of the Young Avant-Garde Ashbery, beginning with these lines:

Old heavens, you used to tweak above us,
Standing like rain whenever a salvo . . . Old heavens,
You lying there above the old, but not ruined, fort,
Can you hear, there, what I am saying?

For it is you I am parodying,
Your invisible denials. [. . .] (*Selected* 71, MSO 202–3)

As the insistent repetition of "old heavens" enforces, the speaker fervently bucks tradition and assumes an oppositional, parodic stance. The abrupt syntactic break after "salvo" suggests a burst of fire against the tweaking admonishments of the past. As the passage continues, the speaker tells us explicitly of his vanguard ambition:

Yet I shall never return to the past, that attic,
Its sailboats are perhaps more beautiful than these, these I am leaning
    against,
Spangled with diamonds and orange and purple stains,
Bearing me once again in quest of the unknown. [. . .]
[. . . .]
But once more, office desks, radiators—No! That is behind me.
No more dullness, only movies and love and laughter, sex and fun.
                                        (*Selected* 72, MSO 204)

These lines are part reminiscence and part reenactment, pitting the old and new against each other with deliberate melodrama, as the oratorical "I shall" and the repetition of "these, these" underscore. They stage an adolescent fantasy of striking out "once again in quest of the unknown" while renouncing conformity and boredom. As a scene in an oblique poetic autobiography, this renunciation of the "attic" of the past and its beautiful vessels, and rejection of the "office desks" of the academy and the business world, suggest the familiar terms of midcentury poetic opposition to verse under the sway of the New Criticism. But to read this passage as a clear window on that moment, on that scene of avant-garde beginnings, one must ignore the forced exuberance of "No! That is behind me." The speaker of this poem knows that in declaring that there will be "only movies and love and laughter, sex and fun," the "only" indicates that this faith too is behind him. The poem dramatizes a youthful aspiration for freedom and the unknown, but it also looks back (from a point less than a decade later) to memorialize a past of which this forward-looking aspiration was a part.

Immediately preceding this defiant apostrophe to the old heavens—right before the passage Ashbery chose for his *Selected*—is a strange museum scene that complicates our reading of Ashbery's early career in terms of an opposition between tradition and innovation. Visiting the old fort, the speaker stumbles upon a museum and responds with this outburst:

> But war's savagery. . . . Even the most patient scholar, now
> Could hardly reconstruct the old fort exactly as it was.
> That trees continue to wave over it. That there is also a small museum
>   somewhere inside.
> That the history of costume is no less fascinating than the history of
>   great migrations.
> I'd like to bugger you all up,
> Deliberately falsify all your old suck-ass notions
> Of how chivalry is being lived. What goes on in beehives.
> But the whole filthy mess, misunderstandings included,
> Problems about the tunic button etc. How much of any one person is
>   there. (MSO 202)

The speaker is immediately skeptical of the museum aspiration to reconstruct the historical past "exactly as it was"—this is not an art museum but a former military site. Using sentence fragments in nonsequiturial apposition, Ashbery offers the image of the trees as a version of Shelleyan leveling sands, and then observes that the museum presents "the history of costume" as "no less fascinating" than the history of political and social upheaval. The observation prompts a string of curses as the speaker wishes to "deliberately falsify all your old suck-ass notions," notions about the "whole filthy mess" the scene seems to embody. Insofar as this slippery passage can be read as a generalized comment, it suggests that individual histories ("how much of any one person is there") have been suppressed by the tidy presentation (exhibits akin to "what goes on in beehives") of matters of archaeological accuracy and chronology ("problems about the tunic button etc"). But this is a stretch. Ashbery presents no coherent critique, but a node of tension and confusion, a tongue-in-cheek burst of indignation and sputtered "misunderstandings."

As if to try again after this inarticulate rant, the poem proceeds to survey the museum at the old fort with regained composure:

Still, after bananas and spoonbread in the shadow of the old walls
It is cooling to return under the eaves in the shower
That probably fell while we were inside, examining bowknots,
Old light-bulb sockets, places where the whitewash had begun to flake
With here and there an old map or illustration. Here's one for
   instance—
Looks like a weather map . . . or a coiled bit of wallpaper with a design
Of faded hollyhocks, or abstract fruit and gumdrops in chains. (MSO 202)

For all of its enshrinement of "old suck-ass notions," the museum still intrigues the speaker with its curiosities—bowknots, light-bulb sockets, flaking whitewash. Its contents are suddenly aestheticized. The exhibits become colorful and suggestive, as the sensory richness of these lines conveys: gustatory pleasures (bananas and spoonbread) lead to acoustic enjoyments (bowknot / socket / hollyhocks; light / whitewash; whitewash / weather / wallpaper). Despite the speaker's antipathy for the old, the museum invites ekphrasis—an effort to describe an "old map or illustration" as having a pattern of "faded hollyhocks" or "abstract fruit and gumdrops in chains." Even a military museum proves to be a fertile source for accumulating poetic material, an impetus for a wide range of verbal responses. The rapid shift in this passage from "war's savagery" to "gumdrops in chains," from indignation to surreal metaphorizing, suggests an ambivalence at the heart of this poem's project of accretion and verbal play. The speaker both shuns and delights in the museum, finding the trinkets and debris of "the elaborate past" to be both repulsive and intriguing.

   Recognizing the equivocation of this museum scene, we better understand why "The Skaters" is a poem in which the vanguardist wish coexists with personal historiography, with a museal project of collection or preservation. The poem opens with the question "But how much survives? How much of any one of us survives?"—the same anxiety that surfaces in the scene in the fort ("How much of any one person is there.") Questions of conservation—not conservatism—work in tandem with the aspiration to innovate. The opening section of the poem describes childhood collections of "stamps of the colonies / With greasy cancellation marks, mauve, magenta and chocolate, / Or funny-looking dogs we'd see in the street, or bright remarks" (MSO 195). These details, culled from a 1911 book entitled *Three Hundred Things a Bright Boy Can Do* (Lehman, *Last* 121), suggest one extreme of a project of gathering and exhibiting, transient juvenile collections of the exotic or unusual (the "funny-looking dogs" are presumably sighted, not stuffed). At the other extreme is the scene beneath the "old heavens" in which the adolescent speaker visits the "old, but not ruined, fort," a scene that invokes the "whole filthy mess" of the presentation of history. "The Skaters" alternates between these two poles of collection—the intimate and the touristic, the child's stash and the state's spectacle.

Somewhere in between is the poetic enterprise of collecting remarks, imagery, anecdote, and allusion, and the attendant matter of collecting and incorporating material from the poetic past and the dominant poetic culture. The pages that follow the scene in the old fort in "The Skaters" are a sorting out of poetic sources. After the belligerent rebuke of the "old heavens," the poem resumes with a voice-over from Eliot: "I call to you there, but I do not think that you will answer me" (MSO 203). We hear echoes of *The Waste Land* moments later: "I heard a girl say this once, and cried, and brought her fresh fruit and fishes [. . .]" (MSO 204). The speaker then nods to Stevens: "I am condemned to drum my fingers / On the closed lid of this piano, this tedious planet, earth / As it winks to you through the aspiring, growing distances, / A last spark before the night" (MSO 203). Then the unmistakable voice of Elizabeth Bishop chimes in: "Suddenly, one morning, the little train arrives in the station, but oh, so big / It is!" (MSO 203), and again, discussing travel, "Is it right? This continual changing back and forth? / Laughter and tears and so on?" (MSO 204).[15] The speaker rejects the "attic" of the past in a classic avant-garde pose, but he also incorporates references to these undeniably "mainstream" or "establishment" sources. The modernist tradition is included in "The Skaters" among allusions to "other" traditions,[16] not as evidence that the poem continues this tradition unbroken, but as part of an allusive omnivorousness that does not respect the skirmish lines of postwar poetic canons.

Thus when a speaker in "The Skaters" insists that he sets out for uncharted poetic waters, his vanguard ambition is necessarily weighed down and complicated by the accumulated exhibits of a collecting sensibility. Directly confronting "the elaborate past" in a museum scene, "The Skaters" takes an approach to poetic collection that is propelled forward but that continues to look over its shoulder at the "old notions" it seeks to "deliberately falsify." When Ashbery republishes his first five books of poems in one volume in 1997, he chooses a title that reflects this tension—"the mooring of starting out." This phrase, from "Soonest Mended," is not the term we would expect for a volume that represents his avant-garde beginnings: "mooring," the dictionary tells us, signifies tying up, not setting forth, as well as a *device* for such securing, an 'established practice or stabilizing influence, anchorage.' The youthful speaker of "The Skaters" prefers the new sailboats to those of the past, but even these new vessels are already moored. In titling

The Mooring of Starting Out, Ashbery brackets his early career with the paradox of setting forth by taking a position—an avant-garde position, but still a position that fixes the movement it propels.

The ambivalence of this early text reveals the complex ways in which Ashbery's avant-gardism has always been interlaced with doubt. Its overlapping layers of irritability, exuberance, and allusiveness reflect a difficult engagement with the terms of cultural codification and vanguard ambition that inform his early—and refreshingly original—poetic stance. The "outermost brink" on which he, writing a few years later, remembered himself poised as a young member of the "invisible avant-garde" is less an isolated promontory than an old ("but not ruined") fort—a scene of visitation of the "elaborate past" that demands both adamant refusal and perceptive openness. The rhetorical occasion of the Yale speech called for a different portrait, a portrait of the artist making it new without the burdening sense of making it new again. It is a portrait, I suggest, that we should not invest with too much weight in Ashbery's career. As I turn to his most famous work, the one that shifts him from outermost brink to literary center stage, I suggest that the right passage to cite from "The Invisible Avant-Garde" is this one:

> The doubt element in Pollock [. . .] keeps his work alive for us. Even
> though he has been accepted now by practically everybody from *Life* on
> down, or on up, his work remains unresolved. It has not congealed into
> masterpieces. In spite of public acceptance the doubt is there—maybe
> the acceptance is there because of the doubt, the vulnerability which
> makes it possible to love the work. (RS 391)

The "doubt element" in Pollock's paintings, Ashbery suggests,—the volatility and undecidability of their connotations, as well as the skepticism with which we may approach Pollock's method and intent—sustains interpretive attempts and repeated readings. This provocation of a critical stance is, in turn, intrinsic to the paintings' aesthetic appeal: we "love the work" because we are drawn into its vulnerability and irresolution.

"CONFUSING ISSUES": REREADING "SELF-PORTRAIT IN A CONVEX MIRROR"

In spite of its status as a masterpiece, Ashbery's "Self-Portrait in a Convex Mirror" remains much more unresolved, much more fraught with doubt and vulnerability, than its celebrants and detractors have allowed. First published in *Poetry* in August 1974, it was quickly applauded by the mainstream liter-

ary establishment, which awarded the eponymous volume the National
Book Critics Circle Award in 1975, and the Pulitzer Prize and National
Book Award in 1976. Hailed by mainstream critics as a culminating moment
in Ashbery's career, it has drawn equally hyperbolic reaction from advocates
of the postmodern canon, who see it as a betrayal of avant-garde aims. James
Heffernan, placing Ashbery in an ekphrastic tradition that begins with
Homer, claims that no discussion of the visual arts in Ashbery's work "could
pretend to be anything more than a prelude to the main event: an explication
of 'Self-Portrait in a Convex Mirror'" (170). Charles Altieri, alluding to the
critical narrative in which the poem marks Ashbery's capitulation to the
mainstream and his abdication of experimental principles, makes the same
point from the opposite position: "One simply cannot ignore 'Self-Portrait
in a Convex Mirror' because even if it is more traditional in focus than most
Ashbery poems, it stands as probably the greatest American poem since the
work of late Stevens" (*Self* 151).[17] In both cases, the poem's exemplarity
hinges on its relation to "tradition," its place in and seeming acquiescence to
a continuous cultural inheritance.

    Yet commentators have overlooked the ways this poem thematizes and
interrogates the very notion of what constitutes "tradition" and an artist's
position in it. By turning to the museum in "Self-Portrait in a Convex Mirror," and "Self-Portrait in a Convex Mirror" in the museum, I aim to challenge an assumption upon which almost all readings of this poem are
based—the belief that because of its more "traditional" posture, it "unfolds
with greater discursive consistency than anything else in Ashbery's oeuvre"
(Lolordo 752). Both praise and censure recur to this sense of the poem's
greater readability relative to Ashbery's other work. I would claim that while
the poem is comprehensible in predictable ways—description and exposition, accessible similes, moments of lyrical intensity—it is also indeterminate
in unpredictable ways. Despite the relative clarity of its opening section, the
poem quickly becomes a tangle of contradiction and indecision, a testament
to cognitive and perceptual ambivalence and attempts to mediate that
ambivalence through an "inexhaustible," ongoing ekphrasis. As the poem
proceeds, and especially by the time we reach the sixth and final section,
which at 242 lines is more than twice the length of any of the others, it is
inextricably enmeshed in the problems of the museum setting in which it
finds itself,[18] problems of homage and resistance, aesthetic autonomy and
iconicity, the elaborate past and the urgency of artistic renewal.

    The uncertainties and vulnerabilities of this poem, its unease about "tra-
dition" and what such a vague term might mean, have been falsely clarified by

its own packaging as a masterpiece. Even a glance at the Penguin paperback edition reminds us that we are in the presence of a superlative literary achievement that pays homage to the art of the past—Ashbery's name in capitals across the top, Parmigianino's portrait in the center, and the three major prizes across the bottom. The extreme instance is the 1984 Arion Press edition, in which the poem quite literally becomes a museum piece. For this limited edition (175 copies), the poem was printed on large card-stock circles, the lines rotating around the center of each page like spokes of a wheel. The circular pages are stacked in a metal canister with original prints by Elaine de Kooning, Willem de Kooning, Jim Dine, Richard Avedon, Jane Freilicher, Alex Katz, R. B. Kitaj, and Larry Rivers. Describing this "apparatus of idolatry," Heffernan observes that "nothing could be more iconic or more hieratic than this way of publishing Ashbery's poem" (171). The Arion edition builds a crypt, a time capsule, in which to deposit Ashbery's work in the archives of the high culture (the canister appears designed to withstand a nuclear blast). Packaged as an *objet d'art,* surrounded by the aura of illustrious names, the poem takes its place in the "special collection."

The Arion edition makes an explicit case for Ashbery's position in a traditional literary canon. The canister includes a recording of Ashbery reading the poem, and printed on the jacket cover is an essay by Helen Vendler that describes the poem's continuity with the heritage of English verse: "This poem is not only an homage to the past of Italian painting; it is equally an homage to the past of Romantic poetry: its waking dream and its ache are a continuation of Keats's own. [. . .] Ashbery lives in the afterlight of Romanticism (chiefly Keats) and Modernism (chiefly Eliot, Stevens, and Crane). [. . .] His emphasis on distortion in the service of expressivity is contemporary, and derives from Stevens."[19] Vendler describes the poem strictly in terms of its relation to the literary and art-historical past. She glosses over Ashbery's experimentation ("emphasis on distortion") by pressing it into "the service of expressivity" and legitimizing it as inherited from Stevens. Denying the poem any attitude of confrontation or even innovative synthesis, Vendler stresses that it "derives" directly from canonical sources, naming Keats, Herbert, Sir Philip Sidney, Hopkins, Dickinson, Eliot, Stevens, Crane, and Wordsworth in the space of six pages. Reading it as a tribute to the "collective Western past," she goes on to make the extreme claim that the poem, "by making Parmigianino, Keats, and Ashbery one artist, makes Italy, England, and America one place; it also makes painting and writing one art, if only for the charmed moment in which tradition and the present, Europe and America, intersect in the musical reverie of language." Vendler's essay, as the general use of the term "tradition" reminds us, universalizes Ashbery's text in a hyperbolic flourish of quasi-mystical ("charmed," "reverie") appreciation.

Yet even as the Arion edition offers an egregious example of the ways a work can "congeal into a masterpiece"—the poem frozen in steel and promotional veneration—it also presents counter-gestures of reluctance and resistance. In the 1983 Avedon photograph included in the case, Ashbery looks a little offended, eyebrows raised and mouth held in a slight sneer. His own foreword to the edition offers a countertext to Vendler's, insisting on the sui generis, the inspirational, and the contemporary, rather than the allusive aspects of the poem's composition. The text that inspired him was not the Norton anthology, as Vendler might have us believe, but an old copy of *The New York Times Book Review,* in which he saw a reproduction of Parmigianino's painting "accompanying a review of Sydney Freedberg's monograph on the painter" in 1950. Ashbery explains that after he saw this newspaper image (a layer of representation that most accounts of the poem forget), he saw the painting itself in the Kunsthistorisches Museum in Vienna in 1959. Then, "one day" in 1973 in Provincetown, he passed a bookstore that was displaying "an inexpensive portfolio of Parmigianino's work," which he bought and "slowly began to write a poem about it, or off it. [. . .] Oddly, when I was in Provincetown the following year I tried to find the bookshop again, but it seemed to have vanished without a trace, like the pharmacy where De Quincey bought his first dose of laudanum and which he was unable to find again." If we must have a Romantic, Ashbery seems to say, then make it the addicted essayist as a model for peripatetic accident. In a deliberate *erasure* of sources, Ashbery makes the poem's antecedents a biographical mirage. The rest of the foreword is an even more explicit, though polite, reaction to the ways his poem has been co-opted by the literary establishment:

> I am of course happy that the poem has attracted considerable attention (of which this new edition is the latest sign), but also surprised. Some have viewed it as a temporary "return to reason" on my part but I don't think so. I firmly believe in the irrationality—as opposed to incoherence—of poetry ("Poetry must be irrational," wrote Wallace Stevens) and think that under an essayistic veneer, perhaps prompted by the logic of art history, this poem is as irrational as my others. Perhaps not. Perhaps it still has something to teach me about itself. (Arion foreword, unpaginated)

Ashbery registers a complaint against the mainstream canonizers by insisting that the poem is less comprehensible than it seems, a point that I will second.

Two "perhapses" insist on uncertainty, and he reminds us that not even he has total understanding of the "irrational" text he generated.

Ashbery's resistance in the Arion foreword reflects his experience of a predicament he addressed a few years earlier, writing about a large showing of contemporary art at the Whitney ("Art about Art" 1978). Artworks in the show, he writes, seem to "honor [. . .] the museum space that is sheltering them, but at the same time they wear their laurels a bit restlessly, as though nostalgic for the scrappy sixties. [. . .] No sooner has the artist dealt with his reactions to the art of the past than he finds himself being wound on the spool of art history [. . .]" (RS 251). In the Arion edition, Ashbery likewise wears his own laurels restlessly, as if nostalgic for his avant-garde beginnings and wary of being wound inexorably on the spool of literary history. This restlessness surfaces in the works of visual art that the edition includes with his poem. A look at the accompanying prints reveals a different side of this strange object—and it is strange, resembling a hubcap the size of a large pizza—than the one we read through Vendler's conservative appraisal. The prints are "blurbs" of another kind, to be sure, celebrity contributions that mark Ashbery as one of the cognoscenti, but they also represent new work by innovative contemporary artists of his generation—riffs, sketches, provisional juxtapositions, oblique commentaries. Ashbery's reasoning for including these prints is illuminating. He relates that he was at first reluctant to include even a reproduction of Parmigianino's painting on the cover of his book, fearing that it would detract from the poem, but that he ultimately agreed because of the painting's "relative unfamiliarity." He then writes, regarding the Arion project, "This edition with illustrations by artists whose work I feel close to seems to me a good idea for the opposite reason of taking the poem away from itself and amplifying it in ways I had never anticipated." This "opposite reason"—including visual art not as clarification but as "amplification"—suggests an effort to resist the "congealing" power of canonization by keeping interart resonances in play. The concentrated force of the "masterpiece" is diffused, made more collective and intertextual, through the inclusion of the work of diverse hands.

To my mind, these accompanying artworks offer the best introduction to Ashbery's poem, suggesting striking tangents and touchpoints for reframing the poem-about-a-painting they address. Two of the prints, most predictably, are portraits of Ashbery, but like Parmigianino's portrait they foreground the uncertain process of portraiture itself, balancing fidelity to and distortion of their subject. Elaine de Kooning's Ashbery is a field of magnified cross-hatching, the slashes that depict his hair and mustache continuous with the slashes of the background. His image emerges from her medium, made by it rather than accommodated to it. Larry Rivers's picture is more intimate, portraying Ashbery at a typewriter tilted into convexity, the text of Ashbery's poem "Pyrography" superimposed over his face and breaking up into indeterminate characters and gaps at the bottom. Rivers inverts the ekphrastic enterprise by painting a text, paying homage to a scene of writing while also reinscribing a visual rendering of what "Pyrography" calls "the force of colloquial greetings" (HD 8). Alex Katz and R. B. Kitaj also offer portraits, but they depict figures other than Ashbery himself. Both pick up on the poem's preoccupation with Parmigianino's enlarged, foregrounded hand: "As Parmigianino did it, the right hand / Bigger than the head, thrust at the viewer / And swerving easily away, as though to protect / What it advertises" (SP 68). Katz portrays a young woman with a sharp widow's peak, her arm raised as if about to salute, a stalled formal greeting rather than a "shield of a greeting" (82).[20] Kitaj recasts Parmigianino's portrait as a contemporary working-class man in a stocking cap, brow furrowed, a gloved hand over his mouth. The image is one of contained ferocity, a stroking of the beard as a reflex against anger. Jim Dine's woodcut suggests an even more ominous revision of Parmigianino's alert repose and Ashbery's temperate interpretation of it. Dine stresses the claustrophobia and terror implicit in the notion of "life englobed" (69), presenting a man straining out of a black background, shoulders compressed as if straitjacketed. All of these portraits multiply and intensify tonal possibilities, undercutting the complacency of Parmigianino's face, and the clemency of Ashbery's tone, with greater urgency. They set in motion a range of responses that are not contained by Ashbery's text but that intervene at various points in its engagement with a visual artifact.

The final two prints resist the notion of illustration by working more tangentially. Jane Freilicher's still life of flowers is at first glance a non sequitur, a refusal of the assignment, but it actually refers to a moment in the last section of the poem: "And the vase is always full / Because there is only just so much room / And it accommodates everything" (77). Freilicher's irises emerge into over-bright light, the black drawing whited out in the foreground like an overexposed photograph. She illustrates not a scene from Parmigianino, but one of Ashbery's many oblique metaphors, one link in a chain of cascading tenors and vehicles. Similarly slippery in its reference and refusal of reference is Willem de Kooning's sketch, which seems to capture the light on the reflective surface of a mirror as a series of s-curves and ink splashes around a central absence. As if looking into a mirror from beneath, he prints four reversed numerals along one outer edge. The sketch suggests the arbitrariness of signage, as well as the fascination with surface on which

Ashbery dwells. Like the others, De Kooning's contribution works in several directions at once, altering our perceptions of Parmigianino's and Ashbery's works by magnifying isolated aspects of them, in this case the counterintuitive readability of a visual surface. By multiplying commentary in this way, the Arion edition invites us to look again at the poem and the painting, to make multi-leveled comparisons, and to consider the play of differences as well as recurrent themes. Most importantly, all of these drawings suggest provisional efforts rather than congealed masterpieces, placing Ashbery's poem in the company of the ongoing work of artists whose innovations he finds most compelling. All of these works have a vulnerable, unfinished quality, indeed an occasional quality, and they invite us to read Ashbery's poem in the same light.

Despite its traditional source, "Self-Portrait in a Convex Mirror" employs ekphrasis as a mode of speculative looking not unlike these contemporary artists' approaches—provisional, approximate, contradictory, continually self-revising. At first glance Ashbery's subject seems an odd (or reactionary) choice for a postmodern poet committed to processual immediacy and surface effect, since it cedes priority to a representational visual work of the distant past, an Italian Renaissance masterpiece. Ashbery takes the risk of seeming to adopt a custodial relation to the art-historical canon, a relation that some critics point to, accusingly, as the superficial reason for the poem's success. Geoff Ward, for example, suggests that because the poem addresses Renaissance art, it has a "European air" that gave the judges for literary prizes "cultural reassurance" (Rasula 290n). To level this accusation, however, one must ignore Ashbery's repeated insistence, in the poem itself, on the contingency and instability of his approach to this work of the past: "The words are only speculation / (From the Latin *speculum*, mirror): / They seek and cannot find the meaning of the music" (69). This speculative verbal striving for elusive meanings is at the heart of Ashbery's various attempts to describe and evoke the painting:

You will stay on, restive, serene in
Your gesture which is neither embrace nor warning
But which holds something of both in pure
Affirmation that doesn't affirm anything. (70)

[. . .] The whole is stable within
Instability [. . .]. (70)

[. . .] I see in this only the chaos
Of your round mirror which organizes everything [. . .]. (71)

Contradictory statements like these—"restive, serene"; "Affirmation that doesn't affirm"; "stable within / Instability"; "chaos [. . .] which organizes"—are exercises in using words as a pliable medium not tethered to a visual referent. With paradox and semantic instability inbuilt in the ekphrastic enterprise, Ashbery finds a means for a fluid, lateral movement among the painting's traces and signs, and a way to adopt a noncommittal stance with respect to the homage that his treatment of the art of the past seems to imply. In other words, the famous painting may come first, but the words slide on a surface of their own invention.

Ashbery further deflects what I have been calling ekphrastic risk—the risk of retrospective deference—by foregrounding his acute awareness of the different settings through which the ekphrastic attempt unfolds. He takes pains to identify the multiple locations, past and present, that inform his experience of this work of art, mapping the painting's composition, his museum viewing of it, and his later recollections against the backdrops of the cities in which these acts take place:

> The shadow of the city injects its own
> Urgency: Rome where Francesco
> Was at work during the Sack: his inventions
> Amazed the soldiers who burst in on him;
> They decided to spare his life, but he left soon after;
> Vienna where the painting is today, where
> I saw it with Pierre in the summer of 1959; New York
> Where I am now, which is a logarithm
> Of other cities. [. . .] (75)

Heavy caesuras afforded by the colons and semicolons in this passage demarcate the careful stratification of three settings. First Rome: Ashbery presents a feat of artistic innovation in its perilous relation to the external forces of its own contemporaneity: "his inventions / Amazed the soldiers." Creation, he emphasizes, takes place in the "urgency" of its particular time and place. Next Vienna: in a first-person, probably autobiographical statement, Ashbery indicates that he saw the painting with his companion, "Pierre," in the Kunsthistorisches Museum. Bringing the work into his own public and private moment, he contextualizes his viewing of it as simultaneously personal and institutionalized. As in "The Skaters," he positions himself as a museumgoer between companionable memory and national treasury. Finally New York: in the city where his own creative composition continues, he asserts a "logarithmic"—calculably intensifying, exponential—relation to these earlier scenes.

Of these three settings for his ekphrastic project, the museum proves to be the most problematic and to demand the most ambivalent inquiry. It confronts Ashbery with two problems that he mulls over in a complex series of questions and answers as the poem moves toward its protracted conclusion. First, the museum marks the exact location where artistic innovation becomes part of a high-cultural tradition, where original "inventions," however they may initially confront or comply with their circumstances, become relics, exhibits, examples of a culture's accepted forms of expression. Second, the museum isolates a work of art from the circumstances of its creation and from the larger world of its audience, conferring an iconic weight that may intensify its aesthetic effect, even as it enshrines the work as a thing of the past.

The first problem preoccupies the entire poem, only to be exacerbated by the concluding museum scene. As Ashbery presents him, Parmigianino is an artist who occupies a difficult intersection of convention and invention. His self-portrait exemplifies a work of art that lays claim to breaking new ground, as a parenthetical aside attests—"(It is the first mirror portrait)" (74)—even as it draws on a legacy of classical balance. Reflecting on the differences between this work and the Mannerist excesses of Parmigianino's later portraits, Ashbery describes an apparent contradiction between harmony and distortion: "The consonance of the High Renaissance / Is present, though distorted by the mirror. / What is novel is the extreme care in rendering / The velleities of the rounded reflecting surface" (74). Giving these lines an acoustic roundness with internal rhyme (consonance / Renaissance) and alliteration (rendering, rounded, reflecting), Ashbery draws our attention to the painting's copresence of novelty and harmony, distortion and careful technique. He underscores both the precision and the ambiguity of the effort: the picture shows "extreme care in rendering" not exact details but "velleities," an unusual word choice that indicates inclinations, suggestions, wishes rather than facts. He quotes Freedberg's comment that Parmigianino strove "'[. . .] not to examine the subtleties of art / In a detached, scientific spirit: he wished through them / To impart the sense of novelty and amazement to the spectator'" (74). Citing this art critic, Ashbery brackets and turns over the terms of an opposition central to the recent history of his own medium—art versus amazement, (New Critical, formalist) objectivity versus (avant-garde, experimental) novelty. Parmigianino stands as both traditionalist and avant-gardist on the basis of this enterprise.

As the poem proceeds, oscillations between the demands of "novel" discovery and the force of convention become more uncertain. Long, ambiguous passages elaborate these oscillations in discussions that are far from discursively consistent. In the fifth section, the speaker awaits the arrival of the "new," but the new is immediately undercut by "preciosity." The word "new" shifts from a general concept to a weaker adjectival role in the space of a single self-qualifying line:

> But something new is on the way, a new preciosity
> In the wind. Can you stand it,
> Francesco? Are you strong enough for it?
> This wind brings what it knows not, is
> Self-propelled, blind, has no notion
> Of itself. It is inertia that once
> Acknowledged saps all activity, secret or public:
> [. . . .]
> This is its negative side. Its positive side is
> Making you notice life and the stresses
> That only seemed to go away, but now,
> As this new mode questions, are seen to be
> Hastening out of style. If they are to become classics
> They must decide which side they are on.
> Their reticence has undermined
> The urban scenery, made its ambiguities
> Look willful and tired, the games of an old man.
> What we need now is this unlikely
> Challenger pounding on the gates of an amazed
> Castle. (75–6)

The line-break on "preciosity" underscores the rapidity with which the "new" is subsumed as a mode of fastidious refinement or affectation. This recognition provokes aggressive questioning—what is the point of challenging tradition, the speaker seems to ask, if the result either "hasten[s] out of style" or "become[s] classic" itself? As in "The Skaters," a Shelleyan figure (here, the winds of time) answers these questions of artistic endurance pessimistically, but Ashbery modifies the trope with a notion of "inertia," which he draws on in its double sense of indisposition to change and continual motion. The passage becomes muddier as Ashbery considers the defeat of "all activity" that this inertia represents—there is always another "new mode" that questions the previous one. Even a strategy of dwelling on "ambiguities," Ashbery's own characteristic strategy (and these lines certainly qualify), eventually becomes conventional, "the games of an old man." Turning to Parmigianino as an "unlikely / Challenger," Ashbery maintains poetic inertia by relying on a seemingly inexhaustible capacity for verbal ambiguity. The passage dissolves into vague pronouns (the word "it" appears 11 times in the next 12 lines), obscuring the contradictions these ruminations have exposed and severing any clear reference to the painting.

The problem of "becoming classic" is even more complicated and equivocal in the discursive final section of the poem, where Ashbery returns to the poignant effect of Parmigianino's face, and ultimately to the museum, to meditate on the ways an artistic enterprise is absorbed into the "collective past." The project of copying an image in a convex mirror, "Though only exercise or tactic, [. . .] carries / The momentum of a conviction that had been building" (76–77). "Conviction" suggests both firm belief and certainty of guilt, as if this "exercise" belonged to the irrevocable past as soon as it was performed, as the past perfect tense in "had been building" emphasizes. A flourish of unattributable metaphor distracts us— "[. . .] the climate of sighs flung across our world, / A cloth over a birdcage"—but the matter at hand resumes: "But it is certain that / What is beautiful seems so only in relation to a specific / Life, experienced or not, channeled into some form / Steeped in the nostalgia of a collective past" (77). The speaker declares impersonally that "it is certain" that beauty, once it assumes form and particularity, inevitably reflects the "nostalgic" terms of a shared history, but moments later a more conversational voice questions this certainty, the cognitive flow leading back to a tension between creation and constraint:

> So as to create something new
> For itself, that there is no other way,
> That the history of creation proceeds according to
> Stringent laws, and that things
> Do get done in this way, but never the things
> We set out to accomplish and wanted so desperately
> To see come into being. (80)

With this resigned colloquialism, the mood shifts into regret about the process by which innovative efforts succumb to "stringent laws" and appear as "convention." Parmigianino's effort to "copy all he saw"

> Remains a frozen gesture of welcome etched
> On the air materializing behind it,
> A convention. And we have really
> No time for these, except to use them

For kindling. The sooner they are burnt up
The better for the roles we have to play. (82)

The painting suggests a "frozen gesture," a sense of inert inscription enforced
by the terse off-rhyme of "etched" and "it." Resisting this feeling of formal
closure by employing looser idiomatic syntax in the next sentences ("the
sooner [. . .] the better," "we have really / no time for these [. . .]"), Ashbery hedges and recurs to the avant-garde desire to burn conventions and use
them only as "kindling" for a renewed artistic fire. Setting these terms in
motion—classic, convention, new mode, style, creation—he toys with various "roles" or attitudes: art lover, critic, master artist, avant-garde challenger.
    Moving among these roles, the speaker turns to the museum setting
that frames them:

Yet the "poetic," straw-colored space
Of the long corridor that leads back to the painting,
Its darkening opposite—is this
Some figment of "art," not to be imagined
As real, let alone special? Hasn't it too its lair
In the present we are always escaping from
And falling back into, as the waterwheel of days
Pursues its uneventful, even serene course?
I think it is trying to say it is today
And we must get out of it even as the public
Is pushing through the museum now so as to
Be out by closing time. You can't live there. (78–9)

From the first, we are put on guard—"poetic" and "art" have been put in
quotes. The "long corridor" is both architectural and mental, a re-approach
to the painting that leads the speaker "back" to the matter of "art" itself.
Strong enjambment on "this," "lair," and "as to" gives these ambling self-
answered questions a tautness, a sense that we approach a pressing matter—
how can this "figment of 'art,'" "darkened" by museum space, have meaning
"today" in the "real" and the "present"? The notion of art's "specialness" arises
as the speaker catches the museum setting in the corner of his eye, noticing
the crowds headed for the exit. Observing this "public" motion, he acknowledges matter-of-factly that art occupies a separate sphere: "You can't live
there." The museum closes each night, and life goes on outside. But he is
reluctant to accept this separation of art from the living, to yield to the ways
the mirror portrait seems "To siphon off the life of the studio, deflate / Its

mapped space to enactments, island it" (75). Using the word "island" as a verb suggests the artificial enforcement of aesthetic boundaries: these reflections on art's specialness in its "islanded" state in the museum, surrounded by "enactments" of the "poetic," provoke ambivalence about aesthetic autonomy that Ashbery does not try to resolve.

Shifting attention from the museum setting to the book of reproductions and then returning to the museum, the long concluding section broaches the related subject of the uniqueness and authenticity of the aesthetic object:

> The gray glaze of the past attacks all know-how:
> Secrets of wash and finish that took a lifetime
> To learn and are reduced to the status of
> Black-and-white illustrations in a book where colorplates
> Are rare. That is, all time
> Reduces to no special time. No one
> Alludes to the change; to do so might
> Involve calling attention to oneself
> Which would augment the dread of not getting out
> Before having seen the whole collection
> (Except for the sculptures in the basement:
> They are where they belong). (79)

The declarative syntax and expository tone here lend a misleading assurance—the rhythm of an argument without definitive terms or logic. We cannot be sure what is meant by the distinction between "all time" and "special time," or why such a change would be unspeakable. The meditation has shifted from reluctance to accept the idea of aesthetic autonomy to the opposite desire to preserve the specialness of the original object. What do we lose, Ashbery asks, by seeing the painting in a book, as opposed to seeing it in the museum, in "real life" and "real time," even if it is still cordoned off from life? In the book of "black-and-white illustrations," the painting is shrouded in "the gray glaze of the past," whereas the museum preserves, literally and ritually, "secrets of wash and finish" and reverence for the skill necessary to produce fine surface effects. These lines evoke Benjaminian nostalgia for lost aura in an age of cheap reproductions (Ashbery's term for the book he found in Provincetown is "inexpensive portfolio"), even as the poem as a whole exemplifies the ways the experience of art has been facilitated by what André Malraux called *la musée imaginaire*. From newspaper image (1950) to public museum (1959) to book of reproductions (1973), Ashbery's treatment of

Parmigianino's painting both utilizes and questions the fluid mechanisms of art perception in his own age.[21]

The ensuing self-conscious reflection on the "dread of not getting out / Before having seen the whole collection" is more ambiguous than it first seems. Heffernan glosses over the negation and reads these lines as expressing the predicament of the modern museumgoer "inexorably expelled by the imminence of closing time, dreading to miss any of the collection" (172). But Ashbery leaves the opposite possibility open: the speaker dreads *not* getting out before he has seen the whole museum—or, to rephrase the line in the positive, he hopes to get out before he sees it all. The speaker dreads the *finitude* of the collection. To see the "whole collection" would mean that the tradition is closed, given over to the "gray glaze of the past." The desire here is for there always to be more to see, even if only some obscure and unnoteworthy sculptures in the basement. In this poem that is so often read as a testament to Ashbery's allegiance to the canonical past, he is far more ambivalent about that past than critics allow. This passage instances his continued desire for an "other tradition" that is not exhausted by overuse, a desire to find alternative sources. (The same inclination is voiced in "Grand Galop," another poem in *Self-Portrait in a Convex Mirror,* as a desire to make use of Wyatt and Surrey as poetic precedents rather than more familiar sources like Keats and Wordsworth [SP 19].) Equivocally, without eschewing the art of the past in this poem about a Renaissance masterpiece, Ashbery reasserts a longing for "tradition" as an open field.

As in "Caravaggio and His Followers," this ambivalent meditation on the museum provokes the speaker's uncomfortable awareness not only of himself as an anxious museumgoer, but as an artist with his own designs on a permanent place in the cultural tradition:

> Our time gets to be veiled, compromised
> By the portrait's will to endure. It hints at
> Our own, which we were hoping to keep hidden.
> We don't need paintings or
> Doggerel written by mature poets when
> The explosion is so precise, so fine.
> Is there any point even in acknowledging
> The existence of all that? Does it
> Exist? Certainly the leisure to
> Indulge stately pastimes doesn't,
> Any more. (79)

Observing himself responding strongly to the painting, the speaker backtracks
to question the elitism of his enterprise. He calls attention to the "leisure to /
Indulge stately pastimes," which clearly had to "exist," despite his insistence to
the contrary, for this poem about a painting to be written, and he worries
briefly over the triviality of poetry and painting both. The self-deprecating
reference to "doggerel written by mature poets" deflects the embarrassing
point that the speaker harbors his own "will to endure" by reasserting a less
lofty poetic ambition. Minus its connotations of inferiority, "doggerel"
describes one of Ashbery's characteristic modes, the use of loosely styled and
irregular measures for irreverent effect. Here, he shifts from a meter close to
blank verse, the formal-sounding observation that "our time gets to be veiled,
compromised," to shorter, insouciant bursts—"Is there any point [. . .]?" It is
one tactic for countering the effects of an approach to art that would, as this
passage continues, "confuse issues by means of an investing / Aura" (79)—a
grander approach with a vested interest in the cultural enshrinement of the
"poetic" and the "figment of 'art.'" As a skeptical, self-conscious museumgoer,
Ashbery acknowledges that this charismatic approach is one he both embraces
and eschews.

The result is, by the poem's own admission, *confusion,* but it is a confusion we tend to forget as we read its resounding Stevensian conclusion:
"Here and there, in cold pockets / Of remembrance, whispers out of time"
(83). Stevens is distinctly audible: "Downward to darkness, on extended
wings" (*Collected Poems* 70). (The metrical similarities of these famous lines
are striking. The spondee "cold pock-" parallels "downward," the softening
in "of remembrance" echoes "to darkness," and the strong caesura signals a
brisk iambic finish: "-ance, whispers out of time," "-ess, on extended
wings.") Ashbery's concluding lines, with their breathlessness and apotheosis
of "remembrance," wake the reader up. Critics love to end their essays with
them.[22] But if we have managed to pay attention to the digressive and difficult final movement that has led up to them, we find that "Self-Portrait"
winds down on a different note, a note of assiduous pessimism: "The fertile /
Thought-associations that until now came / So easily, appear no more, or
rarely. Their / Colorings are less intense, washed out / By autumn rains and
winds, spoiled, muddied, / Given back to you [. . .]" (81). Reading the tri-
umphant final lines, we forget that the speaker has just been worrying about
diminished creative powers, about once-abundant associations that have
been diluted and confused. He worries about these "fertile thought-associations" in the exact terms he uses to judge the "investing aura" of the
museum: "Yet we are such creatures of habit that their / Implications are still
around *en permanence,* confusing / Issues" (81). To forget the ways these

implications confuse issues, to forget Ashbery's persistent habit of dramatizing his ambivalence about creative possibility and permanence, is to misread his masterful, maddening poem.

I have stressed that "Self-Portrait in a Convex Mirror," especially as it interrogates its museum context, becomes a protracted, ambivalent, and often vague meditation on aesthetic autonomy and aura, and on the place of artistic discovery within the unstable parameters of a cultural tradition. If we allow for the poem's confusions as a function of its ambivalence about "traditional" sources and "innovative" stances—a refusal to be nailed down by the retrospective deference that the ekphrastic occasion implies (though the nailing down happened anyway)—we can read its irresolution, its vulnerability to vagueness and contradiction, as a productive strategy. Ashbery seems to have found himself with no other choice than to persist in an "inexhaustible" ekphrasis that allows for its own ambiguity and ruminative elaboration, an inertia-as-motion that might also sustain poetic trust:

> Why be unhappy with this arrangement, since
> Dreams prolong us as they are absorbed?
> Something like living occurs, a movement
> Out of the dream into its codification. (73)

If "codification" is inevitable, as Ashbery seems to believe,—the dream of creation must give way to the finished product's entrance into the museum of culture, into a cultural "code"—then we must operate with and under the "conviction" that "dreams prolong us as they are absorbed." Appealing to a strategy that T. J. Clark, discussing abstract expressionism's difficult relation to the art of the past, refers to as the "can't go on, will go on" syndrome, an effort at "ironic or melancholy or decadent continuation" (371), Ashbery avoids having to revile *or* revere the past by continuing to circulate these prolonging dreams in book after book. In a recent volume (using a royal "we" as he is prone to do), he reminds us of what he is up to: "What kind of clucks / do you take us for, anyway? Everyone knows that once something's finished / decay sets in. But we were going to outwit all that [. . .]" (CY 153–4).

Museum scenes in Ashbery's poetry are an index of his ambivalence about the oppositions that foster this compulsion to fend off the possibility of decay, codification, or stasis. Tracing the vagaries of his peripheral vision in these scenes, his awareness of the framing institutions and ideologies of his experience of works of art, we find that we cannot simply celebrate or radicalize his innovative efforts any more than we can suppress or normalize them. In the halls of an enshrined tradition, he is neither wholly reverent nor wholly combative, and he deals with the many moods and valences of his responses to art in a proliferating array of ekphrastic approaches. I began this study with Andreas Huyssen's urging that we move beyond the reductive dualisms of museum critique, and I want to end this chapter by suggesting that Ashbery's museum interrogations make headway of this kind. Huyssen observes that "[T]he modern museum has always been attacked as a symptom of cultural ossification by all those speaking in the name of life and cultural renewal against the dead weight of the past. The recent battle between moderns and postmoderns has only been the latest instance of this *querelle [des anciens et des modernes]* " (13). In his ekphrastic poetry, Ashbery takes us beyond this latest instance: he does not go on and on in search of the new, nor does he go on and on by paying homage to the old—he goes on and on by doubting and dodging this old quarrel between the old and the new. Writing amidst the clamor of the postmodern era, in 1987, he remarks in a poem that "[m]useums then became generous, they live in our breath" (AG 8). He is alluding to a stance that entails neither rapprochement nor rupture, but a difficult persistence made possible by a vital, abiding generosity.

Chapter Three

Daring to Wink: The Museum Comedies of Kenneth Koch and Richard Howard

Kenneth Koch was a poet of avant-garde rupture, and Richard Howard is a poet of literary rapprochement—or so most critical accounts assume. These close contemporaries tend to be placed on opposite ends of a poetic spectrum, representative figures in opposing coteries. In the early fifties, Howard was part of "a circle of poets and readers at Columbia" that included John Hollander and Robert Gottlieb, and whose efforts involved work on *The Columbia Review* and publication in *The Hudson Review, Poetry,* and *The New Yorker.* Howard remembers in an interview that Auden "was a feature in our landscape," as were their teachers, Lionel Trilling and F. W. Dupee, and he stresses that "we were, as I say, very literary" (Interview 39). In these same years, the yet-to-be-named New York School poets challenged an uptown aesthetic with a downtown one, rejecting the dominant literary climate in which, as Koch put it years later, "One hardly dared to wink / Or fool around in any way in poems" (*Great* 310).[1] As Frank O'Hara described them, he and his friends were "non-Academic and indeed non-literary poets in the sense of the American scene at the time [. . .]" (*Collected* 512). They allied themselves with the art world, edited the avant-garde magazine *Locus Solus,* and openly criticized the work of the other crowd. In "Fresh Air" (1956), Koch's tirade against academic verse, the "Strangler" seeks out and murders "makers of comparisons / Between football and life," "students of myth," and poets who address their poems "to personages no longer living / Even in anyone's thoughts [. . .]" (72–3).[2] Howard, in light of this last indictment, had cause to fear. He exemplified the kind of poet Koch opposed—"[c]ivilized, verbally excellent, ironic, cerebral and clearly [a bearer] of the Tradition" (Carroll 204). Fifty years later, these differences continue to shape these poets' receptions. Fans of avant-garde poetry praise

Koch's inventive and exuberant language, his openness to popular culture, and his irreverence. Fans of the lyric tradition praise Howard's eloquence and intellectual depth, his nuanced measures, and his allusive range.

Yet the term "literary" that divides these groups in their own reflections is not a salient marker of difference. The New York School was always literary, despite O'Hara's protestation to the contrary, even "in the sense of the American scene at the time." Their use of English and French literature has been well documented, and Koch's work in particular is steeped in literary sources, one volume invoking "the names and works of Plato, Aristotle, Sophocles, Juvenal, Spenser, Swift, Wordsworth, and Keats" (Spurr 350).[3] Paul Hoover reports that despite the critical commonplace that New York School poetry is "democratic in its sympathies," O'Hara's "The Day Lady Died" recently elicited the same objection that Howard's poetry often does—a student charged that it was full of elitist references ("Gauloises? Bonnard? Brendan Behan?") ("Fables" 24).[4] Meanwhile, the "literary" poets were by no means unaware of the developments of the avant-garde. It is worth remembering that in 1962, the year that Wesleyan published Ashbery's *The Tennis Court Oath,* the series also brought out Howard's first book, *Quantities.* Howard himself included essays on Koch, Ashbery, and O'Hara in *Alone with America* (1969), and his reading of Ashbery offers one of the most prescient critical observations I know. Searching for "an imaginative schema or construct" to elucidate Ashbery's difficulty, he quotes this passage from André Gide:

> I like discovering in a work of art . . . transposed to the scale of the characters, the very subject of that work. Nothing illuminates it better, nothing establishes more surely the proportions of the whole. Thus in certain paintings by Memling or Quentin Metsys, a tiny dark convex mirror reflects the interior of the room where the scene painted occurs. (19)

"Self-Portrait in a Convex Mirror" is still five years away, but Howard happens upon (or offers up) Ashbery's most famous trope: "[. . .] we are on the way to reading Ashbery on the art of poetry, a tiny dark convex mirror indeed" (21). This observation reflects not only Howard's insight, but the closeness with which the two groups were reading and responding to each other. The circles of the midcentury New York poetry scene show considerable overlap.

Koch and Howard were born within four years of each other in Ohio (Koch in 1925 in Cincinnati, Howard in 1929 in Cleveland), received Ivy League educations, and became professors at Columbia University. Both prolific, their respective bodies of work have earned major literary prizes and awards: Koch, who died in July 2002, wrote more than 16 books of poetry; Howard has written 13.[5] Beyond these biographical similarities, they share several poetic preferences—inventive use of traditional prosody and forms, interest in all things French, fondness for the dramatic. Both have produced poetry that is maximal, not minimal, drawing on capacious imaginations to accomplish poetic feats with unabashed theatricality, urbanity, and self-assured wit. In this chapter, I examine one particularly significant common ground between them: both engage the visual arts deeply and variously throughout their long careers, and both employ comedy in these engagements. In their humorous stagings of encounters with the visual arts—what I call their "museum comedies"—Koch and Howard eschew the solemnity and self-importance of so much contemporary poetry, as well as the ceremoniousness of the museum, by "daring to wink" at both poetry and art.[6] These museum poems, I will show, foreground and interrogate the very issues that appear to divide these poets—matters of tradition and contemporaneity, literary and popular culture, modernism and postmodernism. They call into question the critical commonplaces that guide most readings of their work, and show how the persistent "double canon" (Meek 81) in contemporary American poetry—in which postwar poets are affiliated either with the mainstream lyric tradition or the avant-garde—forces us into the false polarization of saying that Howard pays homage to a cultural tradition that Koch subverts. Koch's parodies of avant-garde innovation and its institutionalization offer a humorous and often self-reflexive critique of the ways postmodern artistic and poetic practices become conventional. Howard's dramatizations of the rituals and conversations that frame aesthetic experience ironize his relation to the cultural tradition he memorializes.

REVISITING THE "LIFE ITSELF MUSEUM": KENNETH KOCH'S COMIC HETERODOXY

Recent criticism has enlisted Koch for the avant-garde canon by defining the ways his work is characteristically postmodern. Introducing him in *Postmodern American Poetry: A Norton Anthology*, Hoover focuses on the influence of the French avant-garde tradition (especially Apollinaire, Max Jacob, Pierre Reverdy, Paul Eluard, and René Char), and comments that although Koch later settled on more traditional narrative forms, his early work is "written in a rigorously non-narrative style" (111). Theodore Pelton describes him as "a postmodern English-language poet" who inherited and reacted against Eliotic modernism. David Chinitz, observing that "Koch's sources include anything in poetry from Ovid to Ariosto to Apollinaire, brought into direct contact with everything from Popeye to the peace movement" (314), argues that

> there is a fundamental relationship between Koch's oft-remarked comic genius and his nearly textbook illustration of postmodernism in verse. With his pointed subversion of genre and convention, his intimacy with mass culture, his opposition to totalizing systems, his implicit (and sometimes explicit) critique of literary modernism, his privileging of the signifier, his self-referentiality, his surrealism, and his reliance on chance, fantasy, and indeterminacy, Koch exemplifies with brilliant clarity the postmodern practice of poetry. And the very qualities that make him so conspicuously postmodern are at the same time the constituent elements of Koch's peculiar vein of comedy. (312)

Koch does all of these things in some measure, but one wishes to warn Chinitz that the Strangler probably also goes after makers of "textbook illustrations." Leaving aside the question of whether items in this list, such as "surrealism" and "self-referentiality," are not just as easily labeled "modernist" as "post," I would argue that reading Koch's work as exemplifying "the postmodern practice of poetry" requires critical tunnel vision. To read him as a diehard opponent of literary modernism, for example, one has to ignore this claim for coherence in "The Art of Poetry": "The lyric must be bent / into a more operative form, so that / Fragments of being reflect absolutes" (174). Or, against the view that his comedy is based on postmodern qualities of indeterminacy and subversion, take Koch's own statement about his influences: "The comic, in a poet like O'Hara or Wallace Stevens or Byron, Aristophanes, Shakespeare, Lautréamont, Max Jacob is part of what is most serious for art to get to—ecstasy, unity, freedom, completeness, Dionysiac things" (Interview 52). As for postmodernism itself, he offers this light verse in complaint:

POST-MODERNISM IN BED
Kandinsky, Arp, Valéry, Léger, and Marinetti
Are kicked out of bed.
Then, for a long time, nobody gets back into it. ("In Bed" 252)

Here, Koch casts "postmodernism" as a puritanical ousting of anyone on the wrong side of debates about "totalizing systems,"[7] lamenting its tenure as a period of exile from the field of pleasure.

My aim is not to overstress terms like "absolute," "unity," and "completeness," nor to argue for a Kochian transcendentalism, but to suggest that what Chinitz overlooks, in his eagerness to align Koch with postmodern poetics, are the ways Koch employs comedy to subvert the conventions of postmodernism itself. I begin by considering his comic strategies as part of a collaborative venture in the company of others in his circle. The primary source of New York School comedy, I argue, and of the formal and generic variety of Koch's comedies in particular, is not postmodernism but ambivalence about institutionalization—whether it is the institutionalization of high-modernist poetry, modern art, postmodern art, or postmodern poetry. This ambivalence manifests itself most pertinently as a collective response to the museum—as a nervous but knowing reflex to giggle, smirk, or salute with satire in the hallowed precincts of the visual arts. I then read Koch's early poem "The Artist" (1958) in the particular light of its museum-consciousness, arguing that it parodies the avant-garde imperative to break with convention and escape institutionalization as one strategy that has itself been institutionalized. Finally, I fast-forward to the other end of Koch's long career to show how the short story "Life and its Utensils" (1993) puts forth a retrospective critique of postmodern inclusiveness. These museum comedies demonstrate not an exemplary postmodernism, but a heterodox and at times romantic iconoclasm: Koch targets the orthodoxies of postmodernism with as much wit and vivacity as he challenged those that preceded it.

In a satirical collaboration entitled "How to Proceed in the Arts" (1961), O'Hara and Larry Rivers state the classic paradox: "Youth wants to burn the museums. We are in them—now what?" (*Art* 94). New York School comedy answers this question in its various voices. We are accustomed to identifying literary modernism and mainstream academic verse as these poets' provocations for irreverence and comic critique, but the museum, especially the Museum of Modern Art, looms as large, and with as much high seriousness. To counterbalance that seriousness, Barbara Guest employs ekphrastic-acoustic wordplay, and O'Hara works in a mode of "parodic nostalgia" (the term is David Sweet's)—two strategies, like Ashbery's as we have seen, that reflect efforts to take pleasure in museums and their contents without succumbing to stodginess or mummification. Koch's own answer to O'Hara and Rivers's "now what?" is synthetic, combining Ashbery's ruminative momentum, Guest's verbal gymnastics, and O'Hara's moody satire in a constant adaptation and renewal of genres, styles, and forms.

To tap the innovative energies of modern art while avoiding its museal elevation, Guest uses material from the visual arts in a pliable and playful way, reveling in the aesthetic experiences it affords, translating it into her own medium, and exploiting wordplay as an antidote to convention. Here is the beginning of "The Poetess" (1973), an ekphrasis after Joan Miró's *La Poetesse* (1940):

> A dollop is dolloping
> her a scoop is pursuing
> flee vain ingots   Ho
> coriander darks   thimble blues
> red okays adorn her
> buzz green   circles in flight
> or submergence?   Giddy
> mishaps of blackness make
> stinging clouds what! (78)

With its ebullient wordplay in the spirit of Miró's busy biomorphic circuitry, Guest's ekphrasis is less an appreciative homage than a lighthearted *paragone* in which she attempts to outdo the visual with acoustic effects.[8] Description of the painting's colors ("coriander darks thimble blues") is a means to creating sonic patterns (the "d" and "b" shift from central to initial consonant sounds). Guest gathers sensory impressions of the painting into a condensed collage of verbal fragments, relying on repetition ("dollop" becomes "dolloping," then off-rhymes with "scoop" and picks up the "p" in "pursuing"), and exclamations ("Ho," "what!") to create a "giddy" tone. She approaches a canonized work of art—in this case the work of a modernist master—without conceding that it occupies a sanctified position in relation to her own project. Guest's "poetess" talks back to Miró's, joyously and irreverently.[9] Koch, as we will see, similarly exploits the acoustic potentialities of language as an antidote to sanctification, eliciting comedy through exclamations, alliterative patterns, and outlandish rhymes. In "Hearing," for example, an ekphrasis after one of Jan Brueghel's paintings about the five senses, he too approaches a museum piece with confrontational whimsy, poking fun at the painter's attempt to depict sound. He hears Brueghel the way Guest hears Miró: "there are roars and meows, turkeys and spaniels / Come running to the great piano, which, covered with pearls, / Gives extra, clinking sounds to your delighted ears" (60).

O'Hara's approach to art in museums, similar in its use of ekphrasis to comic effect, entails "nostalgic" parody with an ever-elusive target of critique (Sweet, "Parodic" 375ff). In "On Seeing Larry Rivers' *Washington Crossing the Delaware* at the Museum of Modern Art" (1955), he sets a tone that is both wry and appreciatively renegade: the speaker longs to be "more revolutionary than a nun" (*Collected* 234). This aspiration allows for a broad range of radical behavior, of course, but it remains his wish—the poem celebrates an avant-garde impulse, even as its manifesto spirit quickly slides into the sardonic. Rhymed plurals throughout the poem ("anxieties," "animosities," "spiritualities") trivialize theorization about "the abstract"—about the dominant artistic advances of the day. If we interpret Rivers's painting as a parody of Emanuel Leutze's 1851 painting in the Metropolitan Museum, we might read O'Hara's title as emphasizing the triumph of one museum over another, the Modern over the traditional. But this opposition is immediately destabilized when we remember that Rivers's picture was considered "heretical in an art world that had given up on representation and was bound to consider a patriotic theme as either hopelessly corny or retrograde" (Lehman, *Last* 319).[10] Is the painting parodying Leutze's aristocratic style and his figure's heroic pose, or is it making fun of the assumptions of an art world that expected representationality to be a parody?[11] O'Hara's attitude toward his subject (and his subject's subject) is also hard to pin down. Is the poem celebrating an artistic effort, or is it an ironic indictment of revolutionary impulses, past and present?

When O'Hara specifies in his title that he sees Rivers's painting "at the Museum of Modern Art," he draws attention to the fact that he is seeing its parodic subversiveness institutionalized as an artistic success. As a personal friend of Rivers and other artists, he saw much new art outside of museums—in homes and studios, for sale in galleries—but here he is caught in the awkward position of addressing a "revolutionary" work the museum has already bought. An employee of MoMA beginning in 1951, first at the front desk and later as curator,[12] O'Hara wrote many of his poems literally against the museum timeclock (especially the famous "Lunch Poems"), but he also wrote against the museum timeclock in a figurative sense. The title of this ekphrastic poem, edgier than it first appears, reminds us that if Rivers scandalized the modern art establishment, his parody was neutralized when MoMA accessioned the painting in 1955. O'Hara wrote "On Seeing Larry Rivers' *Washington Crossing the Delaware* at the Museum of Modern Art" late that same year.[13]

Koch's own response to Rivers's painting was a play, *George Washington Crossing the Delaware*. The script, which is full of slapstick action and posturing speeches by historical figures, refuses to settle on an identity as either school pageant or avant-garde theatre. Koch explained, "I wrote [the play] on commission (no fee) from Larry Rivers for his son's high school production; but the auditorium collapsed, there was no place to rehearse, so the play wasn't done by high school students but by adult actors some years later" (Interview 48). Source and satire also collapse: does Koch's "school play" on this patriotic subject respond to a painting that nostalgically or parodically memorializes school plays on the subject? As with Rivers's painting and O'Hara's poem, it is impossible to put your finger on exactly how, or if, the play is being anti-patriotic—it stays on the surface of its historical hoopla. *George Washington Crossing the Delaware* was first produced at the Maidman Playhouse in New York in March 1962, where two months later another play by Koch was staged, this one taking Art, not History, as its subject. *The Construction of Boston,* a collaboration with Niki de Saint-Phalle, Jean Tinguely, and Robert Rauschenberg (directed by Merce Cunningham), assembled stars of the avant-garde art scene in a comedy of civic formation:

> Rauschenberg chose to bring people and weather to Boston; Tinguely, architecture; and Niki de Saint-Phalle, art. [. . .] Rauschenberg furnished a rain machine. Tinguely rented a ton of gray sandstone bricks for the play, and from the time of his first appearance he was occupied with the task of wheeling in bricks and building a wall with them across the proscenium. By the end of the play the wall was seven feet high and completely hid the stage from the audience. Niki de Saint-Phalle brought art to Boston as follows: she entered, with three soldiers, from the audience, and once on stage shot a rifle at a white plaster copy of the Venus de Milo which caused it to bleed paint of different colors. (Koch, *Gold* 48)

What transpires as these artists perform the "construction" of Boston are predictable *nouveau-réaliste* hijinks, but the words they speak tell another story—a parodic *romance.*

In this production with avant-garde visual artists, Koch puts forth a decidedly traditional literary theme: art as compensation for the ills of civilization, as a way to redress the loss of Nature. The play begins not by subverting the conventions of its genre—the epic story of a place's mythic origins—but by indulging them:

> SAM:
>
>   [. . .] before the first man came to pass
> This site and called it "Boston," there was nothing—
> Merely grass and sea: three high hills called Trimountain:
> Beacon, Pemberton, Vernon,
> And the salt sea—wallahhah!
> And the Cove.

HENRY:

And were there nymphs
Inhabiting this grove?
And demigods, and other treasure-trove
Of ancient days? (*Gold* 49–50)

Koch's chorus describes the ex-nihilo creation of Boston like the founding of Rome, complete with original hills, nymphs, and demigods. Back Bay is personified:

Now I feel the sidewalks, clunk!
Slapping down on me, kerplunk!
And I feel the buildings rising
Filled with chairs and advertising
Where was once a boat capsizing [. . .]. (*Gold* 57)

The polysyllabic rhymes and exclamations, like Guest's giddy wordplay, remind us that the play is not to be taken seriously, but the story is ultimately romantic and nostalgic: once Rauschenberg and Tinguely have done their work (bringing human settlement and architecture to Boston), the Chorus laments, "Something sublime is gone: pure nature—[. . .]. It seems such a short time ago we had that here! [. . .] What this town needs is beauty, what Boston needs is art!" (59–60). The play ends when "Tinguely and Rauschenberg kneel to Niki" (63), the art goddess who has restored beauty.[14] The play form allows Koch to celebrate the romantic pleasures of art as Art, with as much grandiosity, high nostalgia, and over-the-top rhetoric as possible, while distancing himself with comedy—it's all an act.[15]

Coming into contact with the art world, for Koch and the other New York School poets, carried the possibility of encountering art not as the inspiring and flamboyant spectacle enacted here, but as static cultural form. Another strategy for deflecting this possibility was poetic collaboration—by refusing a singular point of view, Koch and his co-poet could catalyze new poetic possibilities, even if they found themselves in a museum. One of Koch's collaborations with Ashbery, "Gottlieb's Rainbow," was reportedly written in the garden of the Rodin Museum in Paris, but it takes as its subject neither Rodin nor the abstract expressionist painter Adolph Gottlieb—"Gottlieb" is a Chicago pinball machine manufacturer.[16] The fact that the place of composition survives in the lore surrounding this collaboration suggests that central to the project was self-consciousness about the museum setting. In other words,

Ashbery and Koch seem to say, we may be poets who go to the museums, but we are not constrained by their stuffiness or by their high-cultural contents. Instead, the museum provides an opening for the energetic activity of writing, an exuberant, prolific pursuit.

One result of this pursuit was a collaboration between Koch and Ashbery entitled "Death Paints a Picture," which appeared in *ARTnews* in September 1958 as part of the series "Poets on Painting." In this list poem, each line of which includes the word "statue" and a famous name, a jovial "Death" envisions a panoramic sculpture garden memorializing figures from Agatha Christie to the Sea Hag, Benedict Arnold to Dick Tracy:

> And the statue of George Washington Carver falls on the statue of Sitting
>   Bull.
> It is the dance of the statues! And the rosy-red Betsy Ross statue creeps
>   into the doeskin tent to sleep
> As blue milk gushes from the statue of Bela Bartok in the night of
>   statues. (24)

Unlike the two ekphrastic poems that appear alongside it in this issue of *ARTnews,* one by Louise Bogan on a fifteenth-century fresco, the other by Hayden Carruth on Mondrian, this poem does not explicate a single work of art, but presents a colorful, clowning museum of Ashbery and Koch's devising, mixing cultural registers in a cacophonous display of weeping, vomiting, and toppling. In a characteristic jab at the sobriety of academic verse at the time, Ashbery and Koch write that "The dripping of the Balenciaga statue on the Popeye statue is interrupted by the statue of T. S. Eliot" (24). The interruption is only momentary, and pandemonium continues in this strange "picture" where the Spanish couturier and the cartoon sailor, among others, noisily bring a wide range of cultural reference into the fine arts setting. The poem concludes with these lines:

> In the Albany public gardens the statue of a young girl stands motionless
>   in the falling snow.
> The statue of Porky Pig oinks at her across the vast waste of white.
> He is reading a comic book showing colored pictures of the statues of
>   men and women. (63)

It is tempting to imagine Ashbery penning the image of the young girl, and Koch elbowing him for getting too close to the lyrical repose of a poem like Eliot's "La Figlia Che Piange." Porky Pig's "oink" disrupts the suggestion of statuesque beauty, and the entire assembly of "statues" is absorbed into a comic book.[17]

Later critical accounts tend to suggest that the New York School's opposition to the mainstream poetry of the 1950s occurred from the distance of "outer Bohemia," but the context in which this poem was published reminds us that Ashbery and Koch's avant-garde word games appeared directly alongside the work of poets who exemplified the decade's official verse culture.[18] Like these poets, Koch and Ashbery found inspiration in the visual arts and placed their work in a major art journal: as the editors of *ARTnews* assure us in a note on the series, all of them are "modern poets who are friends of painters and aware of the new trends in art" (*ARTnews* 57.1: 44). "Death Paints a Picture" airs out the ekphrastic occasion with its chaotic velocity and irreverence (how dreary Bogan's rhymed quatrains on Pollaiuolo look by comparison), but Koch, Ashbery, Carruth, and Bogan are identified here together as "modern poets" and art-world insiders who undertake the traditional literary pursuit of writing a poem about art. This proximity to "new trends" on one· hand and institutionalized literary practices on the other—this nearness to the art world's continual absorption of the new—is part of what instigates the New York School poets' comic forays in the first place. The difference Koch insists upon in his various forms is the antidote of comedy—he never forgets, as so many have, the humor of the avant-garde, and he counters the prospect of musealization with what's funny about what's new.

The humor of avant-garde art piques Koch's interest and awakens his parodic instincts most strongly when others—including artists themselves—need to be reminded of it. A comic didacticism appears throughout his work, and chief among the lessons he seeks to impart is that artistic high-seriousness in its many forms exposes its own folly and superficiality. In "The Artist" (62–9), his particular target is the accelerated pace with which postwar visual artists sought new ways to innovate after modernism, and he parodies (and prefigures) a variety of post-abstract-expressionist efforts to outdo the previous generation in scale, induplicability, and shock value. To dramatize and satirize the institutionalization of innovation (pre-empting several of the questions Ashbery would ask in "The Invisible Avant-Garde"), Koch invents a character and documents his megalomaniacal pursuits in journal form. This form, a verbal sketchbook of prose notes, verse, and interpolated materials, allows Koch the comic latitude to make fun of artistic pioneering while partaking in it. In one self-reflexive moment that encapsulates the poem's theme, he winks at his own formal invention: the Artist remarks, "I just found these notes written many years ago. / How seriously I always take myself!" (65). Koch has to wink, because invention itself, and the seriousness with which it is pursued, is the subject here. "The Artist" parodies the postmodern obsession with "subversion of genre and convention" as it comes into being at midcentury, mocking the obligatory gestures of formal rupture with which artists faced the rapid institutional embrace of the latest new forms.

Koch's formal strategy serves the dual function of mirroring the process he is parodying, and circumventing the charge that in refusing to take avant-garde art seriously, he is writing from a philistine or reactionary stance. He was, after all, sympathetic to the avant-garde impulse in the visual arts, which he saw as inspiring revitalizations of poetry that had yielded to controlled nuance. "The Artist" also appeared in the "Poets on Painting" series in *ARTnews* in 1958, next to three Shakespearean sonnets on Utamaro by Parker Tyler, and the formal contrast could not be more striking. "The Artist"'s unusual shape, with asterisks dividing short sections that shift from lyrical apostrophe to conversational prose, announces that while it may find fault with "new trends," it is not conventional. Instead, it is formally varied and resourceful: Koch's later note indicates that "A way to write the poem I found in Stendhal's *Journal Intime*" (*Great* 321), and he adds a further dimension to this form by suggesting the inclusion of taped-in newspaper clippings and photographs. Strategic uses of ellipses convey the chronological gaps of the lazy diarist, and unmetered lines of varying lengths work against the coherence of the poem as a dramatic monologue by highlighting the disjunctures, hesitations, and redundancies in the Artist's voice. Moreover, the form is flexible enough to allow for a series of notional ekphrases that sketch imagined works. In these descriptions, admissions of ekphrastic uncertainty are opportunities for comic aporias: one work's "shape is difficult to describe," for example. The varied verbal surface and relative formlessness of this poem (compared to Tyler's sonnets, for example) allow comedy to fill the gaps, preventing parody from becoming pedantry.

Throughout this journal of creative angst, the Artist's meditations never get far before circling back to his relation to museums. The poem follows the accessioning and deaccessioning of the Artist's works, and two actual museums are mentioned—the Cleveland Museum and the Indianapolis Museum. Both were building collections of modern and contemporary art at this time,[19] but pointedly, for this New York School poet, they are outside New York. Turning to his Midwest origins, Koch undercuts metropolitan snobbery by setting his spoof of cutting-edge art in the not-so-provincial provinces. The speaker muses on the reception of his relatively tame early works—a statue of a "cherrywood avalanche" in Toledo, a "zinc airliner" in the Minneapolis zoo—and then, in stunts that uncannily suggest land art in the later sixties, travels through Ohio, Pennsylvania, and South Dakota installing ever larger monumental sculptures. Koch lambastes the contemporary art scene, and the art world yet to come, by deflating both artistic grandiosity and the high seriousness institutions must maintain to justify the acquisition of such works as "an open field with a few boards in it" (the Artist's early work *Play*) or six dozen packages of "steel cigarettes" (the Artist's next project, commissioned by the "Indianapolis Museum"). Tracing the Artist's self-consciousness as a function of his museum-consciousness, Koch humorously portrays artists' love-hate relationships with museums, and museums' similarly fickle courtship of them.

The poem begins with the Artist's nostalgic reminiscence about his earlier success, *Play*, which he must admit is playful only because the museum says it is: "Children are allowed to come and play in *Play* / By permission of the Cleveland Museum." To actualize the meaning of the work as designated by its title, the museum (not the artist) authorizes participation and repeals the usual prohibition against touching. Although he has already begun working on his "steel cigarettes" project, for which "orders are coming in thick and fast," the Artist is moved to revisit *Play*. He asks for directions to "the Cleveland Museum's monumental area, *Play*," but is told: "Mister, that was torn down a long time ago. You ought to go and see the new thing they have now—*Gun*." The museum, swayed by changes in fashion, again holds the power to control his work, this time by dismantling it and replacing it with "the new thing" whose title has an opposite connotation. Refusing to be disappointed, the artist insists that he is "thrilled beyond expectation!" that his work has been remembered fondly by the man giving directions, and seeks out his deaccessioned work:

> Now I am on the outskirts of town
> And . . . here it is! But it has changed! There are some blue merds lying
> in the field
> And it's not marked *Play* anymore—and here's a calf!
> I'm so happy, I can't tell why!
> Was this how I originally imagined *Play*, but lacked the courage? (63)

The joke is that the work, in its unmarked setting, is what it always was—"an open field with a few boards in it." The difference is that without being marked, despite the addition of some colorful excrement, it is no longer recognizable as art. Even though the Artist attempts to revel in this freedom from the museum as the true realization of his innovative conception, he admits that he is dependent on the institutional marker for his conceptual work to have value: "It would be hard now, though, to sell it to another museum."

After this experience in Cleveland, the Artist abandons his latest effort to innovate, a critique of commercialism in the form of mass-produced "steel cigarettes," and turns instead to projects that the museum cannot easily encompass or own. In this parody of the monumentalizing impulse, which a note tells us was inspired by a work in the Arizona desert by Max Ernst (321),[20] Koch's method is one of expansion to the absurd. The Artist starts with the site-specific work *Bee*, "a sixty-yards-long covering for the elevator shaft opening in the foundry sub-basement / Near my home."[21] The Artist appears to thrive without the museum's authorization, but Koch makes clear that he still upholds its elitism. When the foundry foreman shows the Artist some sketches and watercolors he has done, the Artist says sneeringly, "It seems that he too is an 'artist.'" By putting the word in quotes, Koch shrewdly has the Artist confirm his own caricature. Despite declaring his independence, he still depends on his status as an "artist" and listens for official reactions to his work: when *Bee* is declared by an observer to be "the greatest thing I ever saw," "Cleveland heard too and wants me to come back and reinaugurate *Play.*" But the Artist is preoccupied with ever grander schemes for works the museum can never house. *Campaign* is "a tremendous piece of charcoal" that "would reach to the sixth floor of the Empire State Building." The Manhattan landmark is a gratuitous detail (presumably the piece would reach to the sixth floor of any building) that highlights the Artist's theatrical monomania for the enormous. The effort to innovate has become a single-minded pursuit of undreamt-of, wildly impractical novelty.

This aspiration for ever grander and more shocking works reaches ludicrous proportions in the *Magician of Cincinnati:*

> They are twenty-five tremendous stone staircases, each over six hundred feet high, which will be placed in the Ohio River between Cincinnati and Louisville, Kentucky. All the boats coming down the Ohio River will presumably be smashed up against the immense statues, which are the most recent work of the creator of *Flowers, Bee, Play, Again,* and *Human Use.* Five thousand citizens are thronged on the banks of the Ohio waiting to see the installation of the work, and the crowd is expected to be more than fifteen times its present number before morning. [. . .] The *Magician of Cincinnati,* incidentally, is said to be

absolutely impregnable to destruction of any kind, and will therefore
presumably always be a feature of this part of the Ohio. . . . (66)

Presented in the form of a newspaper clipping, the description of this work
suggests that it conclusively redresses the failure signified by the museum's
destruction of *Play*—the Artist has succeeded in creating a permanent, public
work that obstructs all commerce and assembles the citizenry in rapt attention.
Koch then predicts the advent of Earthworks when the Artist, late in his career,
turns to striating Pennsylvania hills. Predating by a decade actual works on a
vast geographic scale, such as Michael Heizer's *Double Negative* (1969–70) and
Robert Smithson's *Spiral Jetty* (1970) (Fineberg 324, 327), the poem identifies
a trend toward extreme rejection of generic and institutional confines. Koch
recognizes ahead of time that efforts to carve completely new spaces for art can
be blind to the predictability of their own projects—this Artist's works can
only get bigger. The poem concludes when, in his ever-growing aspiration to
outdo himself in scale and impact, the Artist proposes to represent the Pacific
Ocean with "sixteen million tons of blue paint."
    Through this parody of unfettered efforts to exceed the bounds of what
is artistically possible—to break free of all formal and institutional restraints,
to challenge all conventional understandings of art—Koch shows how avant-
garde innovation quickly becomes its own institution. The Artist becomes a
celebrity, a high-cultural icon who has his birthplace honored and his works
ritualized. As he spends more time participating in ribbon-cutting ceremonies than he does creating art, the journal of his inspirations gradually
becomes a scrapbook chronicling his unprecedented success, a grandiose fantasy of artistic triumph that culminates with the headline "GREATEST
ARTISTIC EVENT HINTED BY GOVERNOR." In this comic tale that
begins with the fate of "*Play*" at the hands of museums, Koch dramatizes a
quest for innovation that leads to publicity-seeking destructiveness. "The
Artist" is a tall-tale spoof of the process of canonization and cultural consecration, and the moral of the story, sweetened by humor, is that grandiosity
and high seriousness about art can turn the original avant-garde impulse of
Play—improvisation, pleasure, creation as recreation, all things that Koch
advocates—into its opposite, the very ceremoniousness and predictability it
sought to dispel.

    In "Life and its Utensils," a short-short story from *Hotel Lambosa*
(1993), Koch returns to the museum to parody the postmodern apotheosis
of inclusiveness. Once again staging a comedy in a museum setting, Koch
offers a generous and humorous critique of his own poetics of inclusion. Of

all the methods by which he elicits laughter—boisterous exclamation, polysyllabic rhymes, hyperbole and litotes, rapid shifts from grandiloquence to
corniness, surreal juxtapositions, direct addresses to the reader—the catalog
or list is his most effective device for evoking what Hoover calls the "comic
sublime" ("Fables" 28), a poetry in which an excess of material is always on
the verge of overflowing its spaces. The early tour de force *When the Sun Tries
to Go On*, written in 1953 as a friendly competition with O'Hara's *Second
Avenue*, consists of 100 24-line stanzas that include just about everything in
a random heap of "cosmopolitan lint"—a tuba, a tarantula, *Timon of Athens,*
and so forth (17–20). In this disjunctive early work, which Chinitz reads as
exhibiting a "nascent postmodern inclusiveness" (314) and which critics
often single out as evidence that New York School poetry foreshadows Language writing,[22] Koch employs a process of extravagant, fragmentary cata-
loguing as a challenge to modernist orthodoxies of formal restraint and
imagistic precision. As we saw in "Death Paints a Picture," making a list of
details from different cultural registers allows Koch and Ashbery to tease "T.
S. Eliot." For Koch and other poets who positioned themselves against the
institutionalization of high modernism that Eliot represented for them,
inclusiveness was an important strategy for creating a poetry too messy and
noisy to be read with New Critical discipline. A similar challenge was issued
in the visual arts against modernist opticality and Greenberg's formalism
through efforts to bring in the "extraformal dimensions of experience," a
challenge that Charles Altieri summarizes with Jasper Johns's claim, regarding Rauschenberg, that "he had managed to 'let the world in again'" ("Ashbery" 809).

    In the half century since Koch began this poetics of collection, how-
ever, inclusiveness has itself become an orthodoxy in response to an orthodoxy, a strategy that has been institutionalized in postmodern art and poetry.
Near the end of "Life and its Utensils," Koch imagines an institution called
the "Life Itself Museum": its mission is to bring into art as much ordinary
stuff as possible, and "Let the world in again" is its mantra. Written in the
early 1990s, Koch's story responds to the ways this privileging of "life itself"
became a dominant aesthetic principle in American poetry in the late twentieth century. We see it at work in confessional poetry and its writing work-
shop derivatives as a glorification of everyday angst and the individual ego at
its most domestic and neurotic.[23] We also see it in the work of the avant-
garde, as Language writers and their followers advocate a poetics of "life
itself" of another kind, tending toward what Hoover calls "language realism"—the effort "to erase the distinction between art and the world" by having "Art [. . .] take on the ordinariness of everyday events, as well as its

plenty, without seeking the heroic, the dramatic, or especially the lyric" ("Fables" 21). Koch, who certainly sympathizes with poetry that strives to connect with the actual and the ordinary, nonetheless expresses ambivalence about the direction that postmodern inclusiveness can take once it becomes the "Life Itself Museum." He seems to have seen this trend coming as early as the seventies, issuing a warning to poets in "The Art of Poetry" while restating his position against humorlessness: "Remember your obligation is to write, / And, in writing, to be serious without being solemn, [. . .] / To be inclusive without being asinine [. . .]" (172). In "Life and its Utensils," he elaborates this warning, staging inclusive asininity in order to mock it. Koch is a consistent and intrepid iconoclast:[24] not even his own aesthetic preferences and procedures are off limits to comic critique. The imaginary museum topoi of this quirky prose piece allow him to stand at an ironic distance from postmodernism in poetry and devise a humorous critique of his own (and others') aesthetic of accumulation and exhibition.

As the story opens, the "sun"—here and elsewhere in Koch's work a casually allegorical figure for poetry, an Apollo of exuberant dailiness—plays the role of sign-bearer for the museum, ushering the visitor into an exhibit of household articles reminiscent of Claes Oldenburg's *The Store*:[25]

> "Life and its Utensils" is at the museum. A song runs toward us with its poster. It is the sun! At the museum, the utensils are arranged according to use: here a knife and fork and platter and chopping bowl and a neon sign saying KITCHEN, another saying MANGER. Here are some dark blue paintings—the utensils, it appears, of Waking up. Here are the seams, the pins, the pink roses and yellow cellophane thread that are the utensils of Summer. The weakness of the show is its disorder, its incompleteness, the evident fact that the user of all these utensils is not here. (*Hotel Lambosa* 50)

Utilitarian function and ordinariness are the organizational principles of this imagined show, principles that Koch parodies by reducing them to the banality of kitchen utensils "arranged according to use." The humor hinges on the neon signs, since their "use" in the context of platter and chopping bowl is their role in reminding us to view these objects as art. "Manger" is mock-clever, signaling semiotic play between the French "to eat" and the English word for feeding trough. Koch thus "privileges the signifier," as Chinitz claims in his list of postmodern qualities, but only to poke fun at artful uses of signage. The unimpressed visitor quickly moves on to "some dark blue paintings" that are suggestive rather than literal "utensils," abstract renderings of "Waking up." The shift from the merely presented to the sensuously evocative leads to the notional ekphrasis of a piece called "the utensils of Summer." For this vision of summer as a sundress in production, Koch assembles a colorful collage both sonically and visually: "pink" echoes "pins," and the long "o" in "roses" initiates the cheerful rhyme of "yellow" and "cellophane." More effective than the randomness of the kitchen utensils collection are the whimsical and resonant details of a metaphor of season-making.

The narrator then assumes the voice of a critic: "The weakness of the show is its disorder [. . .]." This sentence is part of the parody—Koch is not calling for order here, but laughing at critical distaste for the haphazard, even as he points to the problem on which an aesthetic of haphazard inclusion founders. Once inclusion of "all these utensils" becomes a principle, its impossibility functions as its own failure, its "incompleteness." The exhibit can never include everything in "life," and the mock-critical visitor laments in particular the absence of a recognizable and coherent subject, a singular "user." Following the interest and impatience of the museumgoer as he experiences the "Life and its Utensils" exhibit, Koch shrugs off postmodern egolessness and reasserts the importance of the "user's" viewpoint, stressing the museumgoer's receptivity:

> When we came out of the museum, nothing was the same. That is, for a couple of seconds. Everything looked like a utensil: the hill, for supporting the sky; the sea, an electric knife for slicing the sunset's cake, etc., etc. (50)

Like Elizabeth Bishop, who remarked that looking at Bosch paintings made the world look like Bosches for a while,[26] Koch's narrator steps out of the museum and finds his vision shaped by the artistic premise of objects as "utensils." Taking this premise to a ridiculous extreme, he sees elements in the landscape as kitchen appliances. The offhand "etc., etc.," and the commonsense recognition that this effect is only momentary underscore the point that appreciation of ordinary objects as art requires the complicity and creativity of the "user." Exposing the cliché that art shapes the way we view reality, Koch suggests that the impact of this artwork is a function of the museumgoer's willingness to accept the extreme position that everything is art.

As the narrator visits a second museum, Koch parodies inclusiveness even more explicitly. This imaginary exhibition first recalls Yves Klein's *Le Vide* (1958), a completely empty white gallery, and then Arman's response to it two years later, *Le Pleine* (1960), in which the same gallery was filled to the ceiling with garbage (Altshuler 192–97):

In another museum, there is a show entitled "Objects: Lost, and Found." A very large blue mirror is just inside the entrance door. Looking at it, one sees into it, as is the case with mirrors, and there is nothing in it at all. That is the Objects-Lost. The rest of the museum is filled, I mean really genuinely filled, with objects found—everything, gloves, engines, canisters, barrels, flowering plants, love manuals, she-goats, backgammon boards, coffins, fat people, seeds, everything that could possibly, ever, in a lifetime or two, be found. The trouble is there is no room to walk in this museum, no way to see, except from a distance, most of these exhibits, and going into and among them, a person becomes one of them him- or herself. (50–1)

Koch offers two postmodern alternatives and makes fun of both: "Objects-Lost" presents the complete vacuity of a selfless, minimal universe, giving us the paradox of a mirror that offers no mimesis, an opaque absence. Against this piece, Koch poses the clutter of "everything," a humorous string of nouns that defy attempts to find an umbrella category for them but continue to provoke pairings. He is responding less to Klein and Arman than to the ways ideas of inclusiveness were taken up over the next several decades. In a 1979 interview with Joseph Beuys, for example, Frans Haks states, "I can imagine an ideal exhibition or collection in which everything was incorporated" (qtd. in McShine 227). In this fictional version of such a museum, Koch shows what "the trouble is" with this idea: despite its ambition to be complete, the exhibition of "everything" does not embrace the objects of life. On the contrary, it either keeps them at a distance that obscures differences, or absorbs the viewer into the indiscriminate "everything."

Koch's story turns from visual to verbal arts when the museumgoer pauses outside the "Word Museum" to reflect on its "many shows." As he and his companion move through the art scene that these museums represent in miniature, the narrator observes that "Prose and her handmaid, Poetry, followed us": the verbal arts follow the visual arts in adopting an inclusive aesthetic of inventory. The museumgoer offers this "song" as commentary:

A railroad terminal scudded with snow
A lovable scene designer's elbow
These will all fall under the knife
Of the Utensils exhibit of life! (51)

Offering two random details, an urban place with a utilitarian function, and a Prufrock-like fixation on an apparently attractive body part, Koch suggests that in such an aesthetic climate any observed scene or detail can be surgically excised from the "all" and exhibited as a "utensil of life." The awkward rhyme (snow / elbow) creates a faux-somber tone in this jingle, which like the poster carried by the "sun" at the story's beginning advertises a less-than-successful exhibition of the random. Koch's tidy tetrameter quatrain gives verse a cameo appearance. With an exclamation that deliberately falls flat, as if the museumgoer were trying to muster enthusiasm he did not feel for these "exhibits of life," the narrator ends his impromptu performance and goes out "onto the museum promontory to smoke a glass of blue air [. . .]" (51).

This view from the museum, where the narrator takes his surreal refreshment, marks a tonal shift from cheerful mockery to bathos:

> Space travel is gone, Leaves of Grass is gone, Bonington is gone, Delft is gone. All that is left on this highway is a couple of trees. Under them is a car, that is, it turns out, a little museum. "Life and its Dashboards" inside. If you're not distracted by this, I am. (51)

The closed exhibits in this perplexing series are linked by the ways they represent various bygone aesthetics of plenitude and enthusiasm for the real. Koch shifts from "space travel" (quintessential American patriotic romance with technology), to "Leaves of Grass" (founding volume for an American poetics of inclusiveness), to Richard Parkes Bonington (popular English Romantic painter and illustrator of Sir Walter Scott), to Delft (metonymy for Vermeer, or fine pottery). We then learn that "Gerard Manley Hopkins is gone and filigree is gone, filigree effects are gone" (52). Any aesthetic that dwells in vivid realism or sumptuous sensory detail, it seems, gives way to the distraction of a stripped down landscape where we find "a little museum" of car parts. The story then unravels into an infinite possibility of museums, most of which are "gone," including the "Power Museum," "Hooks Museum," and the "Museum of Being Permanently Closed." This last example "isn't a museum, it's a mortuary [. . .]." In this Adornian moment, the narrator laments the momentum of artistic fashion that causes this rapid turnover of "exhibits" and "museums," suggesting that the particular mode of inclusiveness dominant in the present moment—the "Utensils" exhibit that levels all things to mundane functionality—has the effect of canceling other more pleasurable modes of inclusion, and thus deadening artistic possibility.

The narrator then opines that "the Life Itself Museum is drifting away." An avid museumgoer, he observes that the aesthetic of inclusion that brought us exhibits like "Life and Its Utensils" seems to be losing strength.

Koch's story dramatizes this shift as happening in the visual arts, where problems with a postmodern aesthetic of "life itself" were apparent even in its early gestures. In a dialogue with Robert Smithson in 1967, for example, Allan Kaprow noted that in order to keep up with avant-garde art, "Museums tend to make increasing concessions to the idea of art and life as being related. What's wrong with their version of this is that they provide canned life, an aestheticized illustration of life. 'Life' in the museum is like making love in a cemetery" (qtd. in McShine 214). Kaprow's simile is perhaps less neutralizing than he intends it to be, but his point, as creator of Happenings, is that museums moved quickly to embrace the very artistic efforts that sought to avoid the museum's insularity. Koch writes a fable of what happens when we install, in Kaprow's terms, "'Life' in the museum," and he too points out the danger of an exhibition of "life": it gives us canned and aestheticized objects without redeeming energy, novelty, or perceptiveness.

The museumgoer leaves the "Life Itself Museum" as a bored tourist seeking new sights:

> You'll admit that it's been quite a day for going to museums! A medium-sized amphitheatre now, filled with car wrecks, would, for this twenty-four-hour period, be sufficient to round things off. A young schoolteacher (Andy?) who lives in these parts is tagging along with us, asking questions: Do you like our country? Which show did you like best? What, according to you, is the most significant for our time? Oh, all of them, Andy, all! But it would be great to go swimming! The Museum of Oceanography is right here. Great sharks pasted to the pinewood, corals hiding beneath the timbers of the floors, mermaids and mermen dancing attendance, and everywhere the mysterious artifice of salt. After this, finally, we go home, to the non-museum of sleep. (52)

The young academic pesters the narrator with questions about the significance of these new exhibits of "our time," but the narrator dismisses his queries and declares them all to be equally significant. The narrator has grown tired of changing artistic fashions and longs for the immediate, sensory gratification of swimming instead of the distanced perceptions of museums. The joke of the "Museum of Oceanography," of course, is that one probably cannot swim there either, but the narrator indulges a fantasy of sensory contact. The concluding sentences make Koch's point. The under-the-sea scenery, with its evocative diction (corals, pinewood, timbers) and surreal, fanciful imagery, renews longing for "mysterious artifice." Koch underscores the importance of the imaginative, not merely inclusive. Art's "mysterious artifice" is decidedly *not* "life itself," Koch seems to say, and thankfully so. As Hoover puts it, "Denied its distinctions as art, art becomes a democracy of attention [. . .]" ("Fables" 21), and here Koch resists such a democracy of attention in favor of focused, sensory, imaginative experience that is demarcated as such, and that ends.

By claiming that Koch is making a case against the postmodern institutionalization of inclusiveness, I do not suggest that he is lobbying for exclusiveness or for art's iconic sanctity, or against poetic "democracy." What he resists as he parodies the leveling of art with "life itself" is the *elevation* of inclusiveness to a universal aesthetic of non-discrimination. Inclusiveness remains fundamental to his capacious comic sensibility, but he reminds us that inclusiveness need not be asinine—it need not cancel the work of selection and careful sensory construction of detail, or dim the luster of artistic surface. In the course of his long career, Koch saw one museum aesthetic replace another: the modernist museum, with its formalist emphasis on aesthetic autonomy, gave way to the postmodern museum, which offered, as one solution to the problem of viewing art as sacred and inviolate, various strategies for including and exhibiting "life itself"—everyday experience, popular culture, the vernacular, appropriated or found material. By reminding us that these strategies can themselves be installed in museums, circumscribed as art with its own set of conventions, Koch shows the inadequacy of attempts to critique the modernist museum's apotheosis of aesthetic autonomy simply by collapsing the distinction between art and the everyday. In upholding that distinction, Koch's comedy is determinedly heterodox, even heretical in its flouting of postmodern taboos against closure, metaphorical resonance, and aura. As it did early in his career, the museum functions as a comic venue for showcasing conventions (including the convention of subverting conventions) and subjecting them to parody. Writing in and against the museum across the modernism-postmodernism divide, continuing to stress pleasure over the assumptions, fashions, and forms of the moment, Koch reclaims the possibilities of "mysterious artifice."

"ONE MORE FLY IN THE AMBER OF HOMAGE": RICHARD HOWARD'S ATAVISTIC POSTMODERNISM

In "Disclaimers," from *Trappings* (1999), Howard offers this parody of museum policy:

> *The Rape of the Sabine Women*, which the artist painted in Rome,
> articulates Rubens's treatment of a favorite classical theme.

> Proud as we are to display this example of Flemish finesse,
> the policy of the Museum is not to be taken amiss:
> we oppose all forms of harassment, and just because we have shown
> this canvas in no way endorses the actions committed therein. (T 11)[27]

Yoking museum prose into three rhymed couplets, Howard catches "the Museum" in the dilemma of its dual social function: as custodian of the art-historical tradition, it is "Proud [. . .] to display this example of Flemish finesse," but as an educational institution, committed to the uplift of the community, it is nervous about the content of this "example," with its depiction of mass violence against women. Avoiding recourse to the word "master-piece," as revisionary art-historical scholarship requires, the Museum carefully designates the work to be an "articulation" of a "classical theme" and a particular historical-national mode. Rhyming "finesse" and "amiss," Howard encapsulates the Museum's concern that confusion might arise over whether Rubens's canvas is an example of great art or reprehensible human behavior. When he picks up the slant-rhyme again in the next line with the word "harassment," a key word for contemporary political and legal disputes over bad behavior, Howard rounds off a micro-satire of institutional anxieties: the Museum, he suggests, is uneasy about how its relation to Tradition will be construed in a climate of suspicion about the values and motives of high art.

"Disclaimers" is the third poem in a volume that derives a large portion of its material, and some of its methods, from the halls of museums. Ten of the 25 poems are ekphrastic, and five of these have italicized epigraphs that offer information about the artwork's medium, dimensions, and ownership, like wall texts in a museum gallery. For example, the epigraph for the opening poem, "Dorothea Tanning's *Cousins*," reads "synthetic fur over cotton stuffing, / wood base, 60 x 25 x 21 inches, 1970." In light of these "curatorial" framing devices in his own work, Howard's disclaimer—in which "the Museum" is concerned that it will be misunderstood and faulted for its display of a major canonical work—betrays some ambivalence about his own careful displays of fine-art finesse. As the recent publication of *Inner Voices: Selected Poems 1963–2003* reminds us, Howard's oeuvre includes many ekphrastic poems, diverse encounters with Bonnard, Rodin, Donatello, Simone Martini, Toulouse-Lautrec, Joseph Cornell, Nadar, Henri Rousseau, Morris Louis, Canaletto, Fuseli, Delacroix, Romney, Lee Krasner, Henri Fantin-Latour, Gerhard Richter, Tom Knechtel, Yasumasa Morimura and others. Howard draws amply from the materials that museums offer, but his humorous impersonation of the voice of "the Museum" reminds us that his

relation to this conservative institution—conservative in the double sense of
preserving works of the past and upholding a celebratory relation to them—
is not simply one of alliance or inspiration. His parody suggests that while he
may go to the museum, he does not necessarily "endorse the actions committed therein," at least not without some critical distance, ironizing gesture, or
wink of comedy.

Like Koch, Howard casts sidelong glances at museums and the art world
throughout his career. Both poets turn to museums and galleries as sources of
imagery and subject matter, and both draw on their wit and prodigious gifts for
comedy to laugh at high culture even as they participate in it. Nonetheless, with
Howard we are more likely to ask this critical question: does he resist the
museum's sacralizing atmosphere and conservative function, or does he take
refuge there? Before I argue that resistance rather than refuge is the case, I situate
Howard's work within ongoing critical debates about what constitutes the
"postmodern" practice of poetry. I argue that contrary to what we might expect
from a poet whose work is steeped in literary and art history, Howard displays
an atavistic postmodernism. I then consider three poems in which he foregrounds the institutions that frame experiences of the visual arts. Howard's preoccupations with the "trappings" of aesthetic experience—the protocols of
exhibition and appreciation, the theatricalities and civilities, the transactions of
cultural capital—reveal a self-reflexive skepticism that ironizes his relation to the
art-historical and literary traditions that provide his sources.

Like Koch's, Howard's poetry is aggregative and anthological, a collection of styles, forms, and voices. Both are archivists who make use of the mis-
cellany of the past and present, from James Thomson's *The Seasons* (Koch) to
anecdotes about coyote urine (Howard), from ottava rima (Koch) to the
rhythms of a telephone call (Howard). Both generate poetic surfaces that are
complexly layered, at times cluttered, with quotation, allusion, and digres-
sion. Both have produced an oeuvre that can be described, as Howard
describes a museum in a poem I will address in a moment, as "an omnivorous package." Why, then, despite the similarities of their maximal poetics,
and despite their nearly identical class backgrounds and academic affilia-
tions, does Howard's poetry more frequently invite the charge of elitism?
Approaching Howard's work, critics and readers are often perplexed about
where to place it in relation to the intellectual and artistic currents of the last
four decades, decades in which various critiques of modernism have cast a
suspicious eye on lavish and artful displays of high culture. As one of
Howard's characters puts it, "I can't tell if Richard is / very forward or just
very backward" (NT 6). When Bin Ramke calls Howard "one of the English

language's slyest of 'post-modern' poets" (127), why postmodern in scare quotes, and why sly? Is Howard "postmodern" or not?

The short answer to why Howard seems more elitist than Koch is that Porky Pig does not oink in his poems. Though Howard stresses that "the notion of entertainment was powerful early on and continues to be a differentiating factor me" (Interview 43), his overwhelming syllabus of literary culture leaves little room for popular entertainments—he does not meet Chinitz's postmodern criterion of "intimacy with mass culture." The recent sequence "The Masters on the Movies" (TC 51–62) is one exception, a shift into popular territory, but even here he addresses film *classics,* as seen from beyond the grave by canonical literary greats including Henry James, Joseph Conrad, and Willa Cather. On the whole, Howard's work seems to dwell in a hyperliterate zone removed from both popular media and the vicissitudes of contemporary life, and his allusive work can be chastening in its demand that we go to his sources. (Koch, by contrast, lets us off the hook from all that homework: "Do not be defeated by the / Feeling that there is too much for you to know. That / Is a myth of the oppressor" ["Some General Instructions" 165]). Chief among Howard's subjects are indeed the everyday pleasures and pains of working, socializing, loving, traveling, sickening and healing, but he usually negotiates this terrain through literary scenarios. When he does venture out of the library and museum, he tends to protect himself with a scholarly lexicon and bemused decorum. In "Writing Off," for example, his observations about graffiti (cleverly called "cacography") lead him to an inscription of the "inenarrable FUCK" (LMR 35). The adjective is characteristic Howard: he addresses street art, but pairs the ubiquitous expletive with a thesaurus substitution for "inexplicable."

Yet Howard's decorous insulation of the f-word in this instance is misleading: sexuality and eroticism are indeed explicated, often graphically, throughout his work, and this openness counterbalances his allusiveness. Howard may not give us the ordinary life of the body with Koch's cartoon egalitarianism ("Kapow!") but he presents everyday life with frankness all the more daring because his depictions of gay culture and relationships depart from the normative American everyday. The realities of his urban environment appear increasingly in his later work, but unlike Koch, the tone of his meditations on contemporary life is not typically insouciant. Elegies for friends who have died of AIDS, enmeshed within complex commentaries on the crisis, give his work a political seriousness and force that Koch's lacks. The problem of Howard's high-cultural allusiveness in relation to the exigencies of contemporary culture is the subject of one of his poems. Immediately following a poem titled "After K452" in *Like Most Revelations* (1994), we find "Culture and Its Misapprehensions I," in which a friend attempts to puzzle out the significance of the foregoing title. She speculates that "K452" is a new AIDS drug from France, a Union Carbide formula, an abortion pill from France, a plane shot down over Korea, or a code for a "dial-a-sex" line (LMR 59–60). It turns out to be a high-culture reference—a Mozart quintet enumerated by Koechel—but in the process of refuting them, Howard foregrounds these other possibilities. As he calls attention to the undecidability of this reference for even his closest readers, he points out the extreme instability of canons of taste (here, classical music) in our interpretive moment, and destabilizes his own enterprise in relation to those canons.

The reason for much of the critical uncertainty about Howard's work in this interpretive moment, I would argue, is that he plays both sides of debates in contemporary poetry about language, voice, and history. Start with language: one way to frame the question of Howard's postmodernism (or not) is to employ Charles Bernstein's distinction and ask whether Howard's poetry is "absorptive" or "anti-absorptive." The dual canon in contemporary poetry conditions us to answer immediately that his work is "absorptive"—that is, he uses language to achieve "the transparency effect," confident that reference, mimesis, and authorial voice can carry the current of narrative or reverie. Indeed, reading Howard's poems, one has a sense of being absorbed or engrossed in his stories (an uncommon experience in contemporary poetry, but one, not incidentally, that Koch's work also affords). When we consider this effort to engage or seduce with language, and remember Howard's mainstream success, we tend to forget that he also exemplifies "anti-absorptive" uses of language in his insistence on surface, artifice, and linguistic self-consciousness. The same references that give Howard's poems the illusion of a seamless recounting of old stories can also be interpreted as assemblages of quotations and found material, with no hope of total coherence. These assemblages are "anti-absorptive" because they offer a "disruptive presence of a legion of references outside the poem" (Bernstein 56). Drawing from Bernstein's list of synonyms for this "anti-absorptive" tendency, we can describe Howard's work as digressive, exaggerated, distracted, interruptive, stylized, baroque, mannered, ironic, camp, diffuse, didactic, theatrical, and amusing (29). "Absorptive" and "anti-absorptive" tendencies are not mutually exclusive, as Bernstein acknowledges (22), but primary allegiance to "anti-absorption" (or related terms like "impermeability," Perloff's "indeterminacy," and Chinitz's "privileging of the signifier") is a shibboleth in contemporary poetry for being "postmodern" or "avant-garde." With his lyric narratives, fluid syntax, and journalistic clarity, Howard does not pass this test, but if his poems are indeed "absorptive," they are supersaturated to the point that their own excess belies their transparency.

Howard's chorus of literary-historical voices also makes it difficult to place his work in relation to later twentieth-century challenges to notions of voice and history. On one hand, no contemporary poet has a more "voice-based" poetics. Howard makes ample use of dramatic monologues, dialogues, and persona poems, a pursuit of theatrical characterization in formal measures that takes Browning as a model.[28] On the other hand, no contemporary poet has created works that so perfectly embody Roland Barthes's formulation of a text as "a multidimensional space in which a variety of writings, none of them original, blend and clash" (146). Howard, who has translated Barthes extensively,[29] seems to take for granted that "voice" is an unstable construct, and that no textual view-point is singular or unitary, either in relation to himself as author, or in relation to the historical individual it represents. When an interviewer asks him whether there were certain poets he had "to struggle against to define [his] own style," Howard exclaims, as if incredulous that the post-structuralist rejection of originality be taken as anything less than axiomatic, "My own style! Who has his own style?" (Interview 46). Howard's voices are always self-consciously appropriated. What he says in one poem about Nadar, whose work holds a powerful attraction for him,[30] could be said of Howard himself: "You will be obscured by a cloud of pos-tures / and a roster of great names" (M 55). For all its Victorian pageantry, the polyvocal collage of postures and great names that comprises his work is a willing and knowing decentering of the poetic self.

The "roster of great names" that Howard compiles, moreover, belies its own "greatness." He explains that what may look to readers like a glorifica-tion of the past—his focus on the nineteenth century—is partly the result of the accidents of an upbringing with access to his grandfather's library: "it was the Victorians—not just the great ones—who were at hand" (Interview 43). In giving us the demigods with the gods, the version of literary history that emerges in Howard's work is one that is always flirting with its own obsoles-cence. Behind many of his poems, one senses the recognition that few people read this stuff anymore, that these are the ghosts of allusions. Jeffery Donald-son explains that Howard's historical figures are often "untitled subjects," as the title of Howard's third book indicates, and not participants in a transcen-dent history or great Tradition:

> Their identities are often uncertain, at least unspecified. Rather they are
> identified, in the titles, by their dates, a particular station in history:

'1801,' '1858,' '1919,' etc. [. . .] They are figures in time, moments in time, or rather, a way of figuring moments in time. [. . .] Howard adopts the dramatic monologue espoused by Browning as a way of considering those very factors which Browning's characters must disregard, repress, or bracket: moments of time and how they are altered by (and within) historical indeterminacy. (177)

Moreover, Howard's figures are "untitled" in the sense of being without claim to cultural stature or rank: "We have, then, a series of untitled figures, who are the subjects of, and to, a past they struggle to inherit" (Donaldson 177). Howard's historicism thus comprises "separate and unassimilable efforts" to compose history in the plural: "For Howard, there is no redeeming history that can be written or finally known; there are only histories [. . .]" (201).[31] His historical ventriloquism—Barthesian "tissues of quotations," parenthetical elaborations, digressions—is more in keeping with postmodern collage than with Victorian monologues. In this way, his use of history is not unlike Susan Howe's work with Dickinson and Thoreau—an exhuming of archives, with the discovered materials subject to an intense pressure of analysis and remaking.[32]

I propose, then, that we enter Howard's poetic collection aware of an atavistic postmodernism, a tendency that comes into particular focus in poems where he foregrounds the institutions that shape encounters with works of visual art. Howard's project strikes us as so different from Koch's, and strangely out of sync with its time, because of his insistence, as he puts it in one poem, that "the post- must feed on the pre-" (LU 46). As Longenbach recalls, Howard once gave a lecture entitled "Ante-Modernism, or the Blue the New Came Out Of" ("Howard" 153). Howard's poetry reflects this impulse to enlist the aid of a more remote ancestor, an earlier type, in order to engage, interrogate, and ironize the oppositions of the present moment. His poems about museums and related art institutions, I will show, serve the function of bringing this strategy into relief.

If the term "atavistic postmodernism" strikes some as oxymoronic, here's a preliminary illustration: in "To Aegidius Cantor," from *The Damages* (1967), Howard engages a postmodern development in the visual arts—the Happenings of the 1960s—by turning to a fifteenth-century heretic. Aegidius Cantor was, the epigraph tells us, "Inculpated for heresy before the Episcopal Court at Cambrai, 1411" (D 27). Carolee Schneemann's "Meat Joy" (1964), in which nearly naked men and women moved orgiastically amidst raw sausage, poultry, and fish,[33] provokes Howard to summon this figure from the distant past as having a kindred sensibility:

Only you would find it easy to believe
    What we are about these days
Or at least these nights, for little that we do
    Is likely to amaze you,
Minister who flung open the doors of your
    Adamite conventicle
And having suffered the high inspiration
    Of the Holy Ghost (the which
Visited you as you lay with the Brethren
    In quivering chastity),
Ran out into the street, "a long way stark
    Naked," wearing on your head
A platter of meat. It must have been a dark
    Village whose urbanity
You ruffled [. . .]. (D 27)

Howard juxtaposes a medieval religious heresy with a twentieth-century artistic heresy—an avant-garde performance in "our town"—in order to meditate on the "holiness" of the flesh and on the definitions of art in the contemporary moment. Cantor's heresy is his suggestion that human flesh, including his own, warrants the same worshipful attitude as the divine incarnation of Christ: "[. . .] The power / of your nakedness, metaphor of the meat / You wore as a kind of crown, / The charismatic carnal emanations / Not only from the salver / But from yourself [. . .]" (28). But Howard suggests that Cantor offended the town's "urbanity" and sense of propriety more than he offended its theology. His erotic and exhibitionistic activities ruffled his Village; the Happenings, in another Village, shocked and then intrigued the New York art scene with performances that constituted heresy against canonical high modernism in the visual arts and a travesty against the decorum of museum art.

    In this poem, one of the earliest in which Howard takes the visual arts as subject, he considers a form of art that stands in direct opposition to the art of the museums. Happenings, by privileging action and bodily immediacy over cultural permanence, produced not a formal exhibit but, in Allan Kaprow's terms, a "multimedia organism extending into the space of daily existence" (qtd. in McShine 213). Kaprow explains that he rejects the museum because it retains the hushed and reverential atmosphere of its origins in the palace and church: "The modern museum, though up-to-date in architectural style [. . .] still has not been able to shake off this aura of quasi-religion and high rank. It enshrines its contents, still demands a worshipful attitude that reflects benignly on the spectators' growing cultivation and status" (qtd. in McShine 213). A Happening was staged to dispel this "worshipful attitude," to present ever-changing contents that could not be enshrined, and to resist the use of art as a bourgeois means to social advancement.[34] Howard approaches the tensions and oppositions of the visual art of his contemporaries—worshipful attitudes and heresies, acceptable behavior and shocking bodily displays—by taking a "backward" tactic, casting back into medieval history for an obscure figure to corroborate contemporary fondness for carnal theatricality. He stands at a distance from the spectacle, but near enough to acknowledge complicity with the common heresy he recognizes in the Happenings and in Cantor: a break with conventionality and institutionalized decorum that calls attention to the flesh as flesh—erotic, vulgar, mortal. Howard ends on a note of dry humor: "Say / It is the mess we live by, / Made into a joy. The meat joy. You know. Thanks."

Attendance at Happenings notwithstanding, Howard continues to be an avid museumgoer, but not, as we might expect, to take refuge from postmodern challenges to the sanctity of high art. On the contrary, his poems about museums form the crux of his interrogation of elitist and ritualistic conceptions of aesthetic experience. In "Lining Up," the opening and title poem of Howard's eighth collection (1984), uneasy awareness of the museum setting frames an ekphrasis of Rodin's *The Burghers of Calais* and provokes a humorous countertext. The poem is set, the epigraph tells us, in "Pasadena: the museum vestibule."[35] The speaker and a companion take refuge from an impending storm in the museum's glass-enclosed foyer and wait in line to be admitted to the collection. The poem begins "Better stay where we are," a truncated sentence in which four of the six syllables are stressed, suggesting a suspension of movement, a fixing of position and attention in this liminal space. Forced to pause in the vestibule, the speaker notices the museum setting itself—its architecture, the flow of pedestrian traffic, and its pre-gallery spaces. He surveys the light-filled entryway: "[. . .] here at least / we have, however odd, what passes / for a roof over our heads, / and even if the walls are nothing / more than glass, they will be nothing less" (LU 3). The glass enclosure, transparent and yet a barrier to the elements, adumbrates a paradox that the museum raises: it appears to give the visitor an unmediated window on the objects it contains, but only by sheltering them in an artificial environment that announces their cultural value.

The glass enclosure enables the speaker to watch the approaching museumgoers, whom he humorously dubs the "burghers of L.A." Punning on Rodin's title, he observes "these citizens coming upon us / in radiant raiment, the motley / of Southern California[.]" He remarks on their brightly dyed synthetic clothing, and on the "nakedness" revealed by a lack thereof, as if appointing himself the guardian of taste at the museum's door, but this amusement quickly gives way to the recognition that he is part of this haphazard and distracted assembly: "our neighbors / add themselves to the straggling / file we stand in, parti-colored lives / clustered, strung-out, singular, alone [. . .]." As the line-break on "straggling" emphasizes, he and his "neighbors" approach the museum with mixed motives and without ceremony:

> [. . .] some
> eager to sample what is promised them,
> others uncertain why they have come—
>   not turning back but turning
> aside, as if reluctant to face
> engagements they suspect are lying
>   in wait for them up ahead [. . .]. (LU 3)

He recognizes a lateral movement ("turning / aside"), an oblique gesture of reluctance about the "promised" cultural treasure, as if the museum visit were a form of social penance or even, as the phrase "lying in wait" suggests, something vaguely ominous. The speaker then reasserts a privileged position in this motley file, pointing out that the other visitors have overlooked a work of art, an "obstacle in their path, avoided and already / behind them: *The Burghers of Calais* [.]" The stage is thus set for Howard to explore the tension that this museum experience exposes between the unwitting public's reluctance and the informed speaker's appreciation of art. Characteristically, to engage this tension, Howard takes two steps back into the past, first to Rodin "a hundred years back / (before there was a museum / in Pasadena)," and then to Rodin's "fellow snob and countryman / Froissart," "(five hundred years before / there was a Pasadena)" (LU 4).

Howard presents this encounter as accidental—stumbling upon a sculpture while running to get out of the rain—but in turning his attention to Rodin he revisits an artist who had a formative influence on his sensibility (Longenbach, "Howard" 154–5). When vandals dynamited Cleveland's copy of Rodin's *The Thinker* in March 1970, the event caused Howard to reflect on the end of an era of "Great Snobbery"—the era of his parents and grandparents. Howard explains that "Rodin [. . .] fulfilled the needs of the class and circumstance in which [he] grew up—[. . .] [a] class which is, today, a has-been, for it has been had—by its own possessions, despoiled by its own spoils, among which few were so proudly carted home as the statues of Auguste Rodin" (*Paper Trail* 191). These are dramatic musings, but they make the point that Rodin represented an early twentieth-century American bourgeois aspiration for cultural legitimacy—"the last word in middle-class taste" (191). Howard remembers that the biggest book in his parents' house had large letters on its spine. announcing "ART" and "RODIN," the two seemingly synonymous. The 1970 act of vandalism thus signified for Howard the "modern" destruction of an earlier bourgeois social and aesthetic ideology: as Longenbach explains, "Howard doesn't want to dismiss Strauss or Rodin (or blow up statues), but he does want to wrench them from this world—make them 'modern'—by uncovering neglected or obscured aspects of them [. . .]. [T]he poems reread the cultural matrix that shaped him as a child" (156). In "Lining Up," a rereading of this cultural matrix takes place through the juxtaposition of three "bourgeois" groups: 1) Rodin's exemplary citizens, the fourteenth-century burghers; 2) those museumgoers, like Howard himself, whose educational and social background prepares them to revere Rodin; and 3) the distracted, latter-day "Burghers of L.A."

Howard's speaker starts out in the second group, paying homage to a work that Rilke, in his monograph on Rodin, called "[t]he most supreme instance of Rodin's power of exalting a past event to the height of the imperishable" (48):

1)   Jean d'Aire offering the keys
           which drag his muscles down to string,
           ecstatic as he moves, Calais saved,
                to a death an hour away;
[. . . .]
3) Jacques de Wiessant striding, neck outstretched
           to let his eyes see *how* it will come
                before the lean flesh can learn;
4) his brother Pierre beckoning—to what?—
           under his crooked elbow he looks
                back to find the angry stars
           knotted into new constellations;
[. . . .]
6) Jean de Fiennes spreading his arms to let
           rags that must once have been finery
                fall open to manifest
           a nakedness fiercely young again . . . [.] (LU 4–5)

Enumerating his descriptions in these concise units, Howard utilizes a formal device that mirrors Rodin's rectilinear array, a composition that forms an "invisible / cube sealing them together" (LU 5). The image of the keys that "drag [Jean d'Aire's] muscles down to string" suggests palpable bodily tension, as does the image of Jacques de Wiessant "striding, neck outstretched," an effect that Howard intensifies by placing the word "outstretched" at the end of an extended line. The interruption "to what?" set off with dashes conveys uncertainty about Pierre's beckoning gesture and prepares us for the resonant image of "the angry stars / knotted into new constellations": the uncertain future contrasts the shaping of history that is seen as he looks back. Howard then focuses on the way Jean de Fiennes's once-fine rags reveal an erotic body all the more poignant in the context of its impending death. The description emphasizes the ways that Rodin's figures suggest restrained fear, evoking the moment when the hostages leave for the English camp, not knowing that the queen will compel King Edward to show clemency. In his correspondence with the "living bourgeois of Calais" who commissioned the sculpture in 1884, Rodin stressed the importance of the bodies in conveying the piece's significance, explaining that despite the roughness of the working model he was sending them, "My nudes are done, which is to say that the bodies beneath the draperies are done [. . .]" (qtd. in Grunfeld 250). By focusing on the bodies beneath the draperies in his description, Howard underscores the pathos of a sacrifice of "nakedness fiercely young."[36]

The speaker is distracted from this meditation on eros and noble sacrifice, however, by the presence of the other museumgoers in the queue. The museum setting invites a question of relevance: how does this work of art pertain to these people? If the sculpture does indeed exalt an "imperishable" event, its meaning should translate into the present: "And maybe they are with us, / always with us, lining up—maybe / we deserve a share of what they know / and don't know . . . [. . .]" (LU 5). Yet the moment he proposes this empathic connection across history, Howard retracts it. The tone of the poem drops jarringly into comedy as he rewrites his ekphrasis as a parallel list of descriptions of random museumgoers, echoing Froissart's phrase for Eustache de Saint-Pierre's words to the English king, "beholde here we sixe" (qtd. in Grunfeld 245), and suggesting we "Take any six":

1) the tall man, for instance, in tight jeans
   and a ginger turtleneck, the one
        cupping his hands in order
   to look straight into the museum—
   is he discovering that to be
        bewitched is not to be saved?

2) Does the black girl—the one behind him
   in unforgivable (and unforgiving)
            cerise stretch-pants know we live
   as ruins among ruins, rendered lovely
   by staring at ourselves in the glass?
[. . . .]
4+5)   [. . .] And the two
   who move so much like ourselves
do they know what we know: that the great
pleasure in life is doing what people say
            you cannot do? At the end
6) comes a fat woman with a tattoo
   on her left wrist—I hear her sighing [. . .]. (5–6)

Howard parodies his ekphrasis of the sculpture by making it a template for wry observations about contemporary life—here is a group of bodies beneath draperies in catalog colors such as "ginger" and "cerise." He re-envisions Rodin's iconic group as a contemporary tableau, focusing again on clothing to make the comparison. In the sculpture, finery reduced to rags conveys the erotic pathos of a noble sacrifice, but in the museum, flashy clothes are anti-heroic signifiers of contemporary disconnection. Looking first at the "tall man," he differentiates between "bewitched" and "saved," resisting the idea that art has redemptive power and suggesting that it can only divert. In the second description, he suggests that narcissism is a function of decadence—that we live as self-absorbed "ruins among ruins." In the third and fourth, more optimistically, he seems to recognize another gay couple, reflecting on the pleasures of defying social conventions and prohibitions ("what people say you cannot do"). The sixth figure sighs wearily, completing the comic comparison: in Rodin's group, the six figures are lining up to give their lives for the good of the community; in the museum foyer, the public is lining up out of a sense of cultural duty (or just to get out of the rain). The heroic, Howard suggests, no longer obtains in the culture this museum serves.

The poem foregrounds a central tension in Howard's work: Rodin's sculpture exalts a past event to the "height of the imperishable," and it represents, for Howard, a decidedly perishable bourgeois class sensibility, the post-Victorian idolatry of Rodin with which he was raised. The speaker begins from a position of elitist reverence, confident that he has the good taste to appreciate art more than his neighbors. In the wake of the comic comparison of an ekphrasis and its countertext, however, the speaker's position with regard to the museum is unsettled. The museum now appears to him as a chilling mausoleum of cultural memory: "in here, now and forever, death / of a kind." The poem concludes with the speaker's self-conscious and ambiguous realization: "I am standing first in line" (LU 6). Is he standing ahead of the others in a position of leadership, or is he, like them, awaiting disappointment, or confusion, or uncertain reprieve? The speaker is no longer sure that he is a guardian of taste or a model citizen of any kind:

> even before we get in,
> futility bears down on fatigue
> in irresponsible foyers where
> > a man can know everything
> but nothing else. The omnivorous
> package waits, and our riches blind us
> > to our poverty . . .
>
> > Bundle up
> against the weather and wait your turn;
> we are standing where the burial-
> > places of our memory
> give up their dead. MUSEUM OPEN
> SUNDAY UNTIL FIVE. ADMISSION FREE. (LU 7)

There is bitterness here, regret that "the omnivorous package" comes down to this inert, perfunctory space where a lavish display of cultural "riches" only masks "futility" and "poverty." The speaker issues a parental command to "bundle up" and "wait your turn," as if demanding respect for "our memory" that no longer exists for himself or for the disarrayed individuals in the crowd. The reproduction of the museum's sign in emphatic capitals reads ironically, reminding us that despite the institution's announced democratic openness, the meanings of its contents are not accessible to all. Weekend hours may permit the "Burghers of L.A." to attend for free, but the museum welcomes a bourgeois constituency for whom not even a central bourgeois value—appreciation of art as exalting the past—ties them to the works they are about to view.

The speaker's uneasy position in this threshold space, waiting to enter the museum but not quite in it, torn between his ekphrastic homage and its comic undoing, generates the sense of ambivalence I have been calling Howard's "atavistic postmodernism," a self-conscious skepticism about the cultural inheritance that the museum represents. This midcareer poem suggests a subtle mode of "postmodernist parody," a mode that Andrew Epstein, drawing on Linda Hutcheon, defines as the "'installing and ironizing' gestures [. . .] which summon up echoes of past texts only to subvert them," gestures that in turn constitute a stance of "critical irony" (121n). The past text in this case is Rodin's sculpture, metonymy for "Art" in a particular bourgeois social milieu on which Howard draws, and he reads it in ironic relation to the contemporary audience for whom it is installed. As awareness of the museum setting brings his "critical irony" into focus, Howard invokes Rodin as a cultural ancestor only to destabilize the project of canonizing the noble dead.

"Even in Paris," from Howard's subsequent book, *No Traveller* (1989), pivots on a comical museum experience in which the character "Richard" resists the ritualization of the aesthetic encounter. Critics tend to consider Howard the most ceremonious of poets, but in this poem he desanctifies a ceremony that takes place in the museum. As Longenbach observes, Howard seldom generalizes about trends in contemporary poetry, but when he does, in an obscure journal six months after the publication of *Alone with America* (1969), he recognizes this common thread: "a mistrust, a questioning, indeed an indictment of all the overt ceremonies which constitute—which always *have* constituted—the means of poetry" (qtd. in Longenbach, "Howard" 142). Longenbach argues that the poet this statement describes *least* is Howard himself, whose work does not constitute a "poetry of forgetting," but quite the contrary. Yet when we observe the behavior of Howard's speakers in museum settings, where they are surrounded by the overt ceremonies of high culture, we cannot be so sure that Howard trusts these rituals as unconditionally as Longenbach claims. In his museum-conscious poems, Howard takes part fully in the postmodern mistrust of aesthetic sanctity and aura that he identifies in his contemporaries.

"Even in Paris" is an epistolary tale of an imagined encounter with Wallace Stevens in Paris in the fifties.[37] Letters from Richard and Ivo to their friend Roderick back home in Schenectady describe meeting the poet, whom they are surprised to discover falling on his knees in the Sainte Chapelle "as if receiving / stigmata from stained glass!" (NT 5). Stevens was, as the title of Howard's volume attests, "no traveller," but he wrote of his desire to visit Paris throughout his life. As he surmised in 1950, five years before his death, "I suppose that if I ever go to Paris the first person I meet will be myself since I have been there in one way or another for so long" (*Letters* 665).[38] Howard convincingly dramatizes the self-revelatory encounter that Stevens anticipated would have taken place if he had not taken his own advice in "Prelude to Objects": a man "has not / To go to the Louvre to behold himself" (194).

If we turn to this advice in its context, however, Stevens's assertion that museumgoing is unnecessary is more ambivalent than it first sounds:

> If he will be heaven after death,
> If, while he lives, he hears himself
> Sounded in music, if the sun,
> Stormer, is the color of a self
> As certainly as night is the color
> Of a self, if, without sentiment,
> He is what he hears and sees and if,
> Without pathos, he feels what he hears
> And sees, being nothing otherwise,
> Having nothing otherwise, he has not
> To go to the Louvre to behold himself. (*Collected Poems* 194)

Stevens starts to sound like Gertrude Stein as he presents these big "ifs." According to these criteria, a visit to the Louvre is unnecessary only if one achieves immortality, a life of sensation without sentiment, and a complete evacuation of the self in unity with perception. If we do not meet these requirements, we must conclude that a trip to the museum is in order. Howard seems to have written "Even in Paris" as a kind of poetic dare: he has "Richard," a version of his younger self, lead Stevens to the Louvre.

The man whom Richard meets and believes to be Stevens has exceptionally high expectations for the museum experience. He wants "to make 'sense' of the *Nymphéas*," Monet's murals in the Orangerie:

> "I have been told one is embraced, they curve
> around one in a continuous ecstasy . . .
>
> It seems worth leaving even Hartford for that.
> [. . . .]
> I'd like to let those water-lilies have
>
> their way with me; I'd like to learn from them:
> if anything could be explained, then everything
> would be explained . . ." [. . .] (NT 10)

Describing his hopes for the experience in sexual terms, Stevens seeks an ecstatic communion with the mural that doubles as an education in "everything." His escort dutifully accompanies him, but is less hopeful: "Well,

dear, we reached the *empty* Orangerie / (day-after-Christmas void) and there
we stood, / enveloped by the ovals of nenuphars" (NT 12). Noticing that the
social environment of the museum is uninteresting, Richard describes the
paintings with deadpan sonic play, linking "enveloped" with "ovals" and the
exotic-sounding French word for waterlilies, "nenuphars." He is less than
impressed by the spectacle, as the color words emphasize: "a cycle of mustard
and mauve / makes it hard to link how much there is *of* it / to how little there
is *to* it" (NT 12). Though he willingly leads Stevens to the site of a ritual
encounter, he himself cannot partake in the ceremony. The ensuing scene is
worth quoting at length:

> [. . .] That's what I saw:
>
> my poet paralyzed by the perimeter
> of a wave without horizon, without shore . . .
> He stood stock still, and I think it was awe
>
> he felt at how the visual could turn
> visionary. He stayed there a long while
> (I, meantime, loaded up on postcards: X
>
> marks where he stood, admonished by Monet.)
> "We also ascend dazzling," is all he said,
> or all I could make out—is it a quote?
>
> You'd have thought I had *awakened* him
> by shouting in his ear; he started up
> when I *whispered* was he happy? "Happy here?
>
> —how hideous the happiness one wants,
> how beautiful the misery one has! . . .
> I think I'll stay a little longer here.
>
> Alone." I left him then, of course—
> mine was the backward glance of Orpheus
> or of Lot's Wife, the unretarding gaze
>
> that loses the beloved where last seen:
> my Sacred Monster loomed, one big black lump
> in a circle of besieging light, and Rod,

> he was slowly, in a sort of demonic shuffle,
> turning, turning round the oval room,
> palms out and humming harshly to himself—
>
> it was, I could tell, a ritual exploit
> danced by the world's most deliberate dervish—not
> whirling but centripetal. [. . .] (NT 12–13)

Telling the tale in Stevensian tercets, Howard positions his speaker as standing aside, narrating an experience of aesthetic intoxication from an ironic distance. He comes to the museum in a spirit of groupie-ism, thrilled to play personal tourguide to Stevens as a poetic celebrity (as the epithet "my poet" emphasizes), and he watches in surprise as Stevens appears to experience the sublime: facing a "wave without horizon" that transforms the visual into the visionary, he is paralyzed by awe. Richard is somewhat incredulous, qualifying the word "awe" with the phrase "I think it was," but Stevens has clearly been transported.

Howard undercuts this account of rapture with a parenthetical aside: Richard, more comfortable as a consumer than an acolyte, runs off to the museum shop to "load up" on postcards. An "X" marks "where he stood" on a cheap reproduction, "admonished" rather than ravished. Unlike Stevens, Richard locates meaning in a literal "X," reducing sublimity to souvenir, and his flustered, gossipy interest provides a counterpoint to the ritual encounter. Overhearing Stevens's mumblings, he then asks a question that Howard's readers are forever asking—"is it a quote?" It is in fact, from section 25 of Whitman's "Song of Myself":

> Dazzling and tremendous how quick the sunrise would kill me,
> If I could not now and always send sunrise out of me.
>
> We also ascend dazzling and tremendous as the sun,
> We found our own my soul in the calm and cool of the daybreak. (50)

In the midst of the humorous account, Howard offers a fine ekphrasis of the mural's effect, suggesting that Stevens is justly experiencing the extremes of Whitmanic vulnerability and omnipotence: "[. . .] It was the play / of surfaces that held him, infinite, / / centerless, unstructured: only ecstasy, / an airless moment when he might not send / the water-lilies back" (25). At the same time, the question "is it a quote?" reminds us that Howard stands at an ironic distance from his speaker (Howard himself would not have to ask).

Richard is an ephebe still finding his way in relation to his poetic forebears, and here in the museum with Stevens, his idolatry breaks down. He cannot muster the visionary enthusiasm of the great high modernist. When he awkwardly asks Stevens if he is happy—asking in other words if he is glad he has come to the museum—Stevens interprets the question as a metaphysical one, and Richard leaves his poetic hero out of politeness and perplexity. He does not have the "aptitudes of worship" that Stevens does. His poet becomes a "Sacred Monster," a "deliberate dervish" whom he watches with embarrassment as he does a "demonic shuffle." In this comic deflation of "a ritual exploit," Howard ironizes his relation to Stevens's transcendent vision and to the museum experience that fosters it.

"Even in Paris," though it is amused and forgiving in its critique, exemplifies the mistrust, questioning, and indictment of "overt ceremonies" that Howard identifies in his postmodern contemporaries. By resurrecting his poetic ancestor and caricaturizing his stance—again enlisting the past to unravel his relation to it—he grapples with the legacy of high modernism he inherited from Stevens, and more directly, from Auden. In one poem about Auden, Howard states that the older poet advised him to "praise" the "sacred places" (FF 15). Longenbach, like most of Howard's critics, argues that under Auden's guidance, Howard "became a poet whose goal was precisely to praise the sacred places and encounters" ("Howard" 145), but this view overlooks Howard's equivocation. In "Even in Paris," Howard does visit one of the "sacred places," but he stands apart from the sacred encounter. The story of meeting Stevens in Paris ends not with profound poetic give and take, but with a comic "epiphany" in which Richard's socialite friend Ivo bumps into Stevens later at the Tour d'Argent—Stevens jumps up in terror when he hears someone call "Wallis" (the Duke and Duchess of Windsor are dining with Ivo) and thinks someone recognizes him. Ivo sums up the poem by explaining the source of Richard's disappointment with his adored precursor: "[. . .] I believe Richard lost his way looking / for a genius who might fuse / life and art—[. . .] / [. . .] / Richard, of course, with his love of poets / second only to his love of vulgarity / *would* revel in the humbug—/ one more fly in the amber of homage" (29). With its climactic scene set in the museum, Howard's outrageous fiction offsets homage to genius with the comic extremity of its ritual scenario.

In his two most recent collections of new poems, *Trappings* (1999) and *Talking Cures* (2002), Howard again employs ekphrasis as a mode of tribute and then ironizes that mode with comic resistance. The five-part series "Family Values" in *Trappings* exemplifies this double stance. The first three poems, serious treatments of paintings of Milton dictating *Paradise Lost* to his daughters, bear conventional ekphrastic epigraphs: "after Fuseli," "after Delacroix," and "after Romney." Howard uses the paintings as starting points for dramatic monologues in the voices of Anne, Deborah, and Mary Milton, a complex sequence of reflections on the reciprocities of the "sister arts," the genderings of loyalty and sacrifice, and the origins of poetic creativity. The fourth poem, "after Mihàly Munkàcsy," begins a slide into the odd and amusing, relating the presumed "crackpot correspondence" between the descendents of the painting's sitters and the New York Public Library. The fifth, as we will see, gives the series its comic relief. All five poems fall under the heading "Family Values," a title that invokes the omnipresent catch phrase of neoconservative politics to put a satirical spin on the dynamics of these various families. More directly, the title replies to Howard's earlier poem "Personal Values." In this poem in *Fellow Feelings* (1976), the speaker writes to Magritte to thank him for the "magic reason" that offers the speaker his "life's illustration" (seizures give him heightened glimpses of objects "obscene / With enormity" and framed against the sky). "Family Values" finds its way back to Magritte by a circuitous path, shifting from "personal" identification with an artistic vision to the social groups and institutions that dictate (and take dictation from) aesthetic experience.

In "Family Values V," a comic exposé of literary "values" as a function of "sin, sales, and Surrealism," Howard writes against the seriousness of the poem's preceding sections. Unlike the ekphrases in the first four "Family Values" poems, this final poem describes a painting that does not exist. Howard invents "a new canvas / by René Magritte" entitled "Family Values," which "Magritte has informed us, is his tribute / to Fuseli's masterpiece (now on view / at the Art Institute of Chicago)" (T 33). Notional ekphrasis becomes a mode of off-color comedy as various details suggest the "obscenity" of this imagined canvas, which apparently depicts Milton as a fish lying on top of three nude daughters with fish heads. The image is one of Howard's own devising, but he may have had in mind several of Magritte's paintings from the mid forties, including the "quasi-impressionist" works *The Age of Pleasure* and *The Fire*. In the first, an enormous toad clings to the long blonde tresses of a naked woman as she touches her nipple; in the second, the same woman and toad appear in the company of two other nudes as the Three Graces—three blushing daughters who infuse literary lore (here, Grimm's "The Frog-Prince") with frank sexuality (Sylvester 260, 264; Whitfield 92). Even more strikingly, Howard may also be appropriating Magritte's *The Old Gunner* (1947) (see figure 2), in which a giant man-fish (penis-like rather than phallic) embraces a nude young girl between his thighs, one of which ends in a peg leg. As Howard taps these associations, Magritte's disturbing combinations of eroticism, classical poses, and amphibious creepiness offer a counterpoint to the heroic invocation of Milton and his amanuenses.

Figure 2. René Magritte, *The Old Gunner,* 1947.

Giving us this weird (and unseen) countervision, Howard interrogates the ways literary and art institutions canonize their subjects through a mixture of professional decorum, commercialism, and voyeurism. The poem imagines a 1947 correspondence between literary scholar Marjorie Nicolson and gallery owner Julien Levy. In an obsequious appeal to "the prestige of [her] immense authority" as the "resident Miltonist of Columbia," Levy solicits Nicolson's endorsement of Magritte's "new painting," which has raised "iconographic questions and even / charges (in Belgium) of obscenity!" Nicolson is "proud to be consulted: rarely / is an academic figure [. . .] / asked to mediate / among sin, sales, and Surrealism." Dramatizing this exchange, Howard gives us a glimpse of the forties literary and art establishment (Meyer Schapiro suggests that Levy contact Nicolson), which turns to rank and expertise to reconcile the prudish public with modern art. Unlike Fuseli's masterpiece on view at the venerable Art Institute of Chicago, Magritte's modern painting on display in a New York gallery requires authoritative commentary so that it "might work its magic / unobstructed by the philistinism / which has so often blighted the careers / of Ernst and Balthus, Bellmer and the rest"—in other words, so that it might be sold. Nicolson's response is both accommodating and self-deprecating. She visits Levy's gallery and makes fun of her own persona, telling him not to address her by her academic title: "'professor' just makes static in my ears / and puts off any kind of intercourse with actual learning. / I much prefer that you would think of me / as Chairman of the English Department." Howard presents only Nicolson's half of the dialogue, but implied is Levy's concerned reply about the gender incorrectness of "Chairman," an objection which Nicolson dismisses: "No, *Chairman* will do. The gender affixed to 'chair' hardly seems / a matter of controversy." (Her students refer to her as "a tweed fireplug with breasts." She rejoins, "Absurd simile: fireplugs, as you know, / already have breasts!"). Through this amusing banter, Howard reminds us that the reception of art is a social process—a function of the institutional and personal connections of a particular time and place. (T 33–5)

Nicolson's interpretation of this new "Magritte" highlights a tension between institutional propriety and artistic license to portray the erotic or obscene, a tension that we find throughout Howard's work when his explicit and often funny bawdiness works against his decorum. Nicolson begins by placing Magritte in an art-historical context even as she mocks the scholarly obsession with movements: "I am assuming that Magritte is still, / if not in favor / with Monsieur Breton, a Surrealist?" She then addresses the question of whether the picture is "immoral" for its apparently incestuous content by appealing to a canonical great: "Shakespeare was hardly / the first to present / a father whose feelings for three daughters / infringe on our notions of seemliness." She ventures the idea that Magritte is punning on "milt" when he portrays the deposition of fish sperm on ("milt on") the daughters:

> an incestuous poetry with his
> > ichthycephalous
> daughters . . . Oh dear, I suppose that does sound
> rather obscene, though it's a classical
> > trope for Apollo. (T 36)

She attempts to confer an official dignity on the painting by assigning the Greek term "ichthycephalous," but it makes the matter worse (in part because of the "ick" sound). Despite her invocation of a classical trope, she cannot dispel the sense of impropriety about these "fishy goings-on": "The longer I look, the less I know what / to think. Magritte's title is certainly, / like many of his, / odd and perhaps perverse." Deftly leading his character into this quandary, Howard underscores the ambiguity and perhaps irreconcilability of matters of conventional taste, official endorsement, and aesthetic judgment about sexually explicit "modern" art.

Howard concludes the poem by foregrounding the institutional trappings of the experience of art. Printed at the bottom of the poem's final page, a box suggests the text of an invitation card, complete with street address, time, and date, for the Julien Levy Gallery's opening of a show "featuring *Family Values*, a new painting." At the bottom of the card is a blurb: "No modern canvas has given me more pleasure.—*Professor Marjorie Nicolson*" (T 37). Nicolson's thoughtful but ultimately befuddled interpretation of the painting is reduced to this hyperbolic cliché. Levy does not need her analysis; he needs the transaction of cultural capital that her endorsement makes possible. To enhance the work's cultural (and, in turn, economic) value, he quotes her praise and includes the title she shunned. Her official role as "Professor" at an Ivy League institution gives her remark its sole meaning, that of decreeing that this "modern" painting is commendable as high culture. Ending with this imagined text published by an art gallery, Howard's series has come a long way—from meditations on Milton's biography to a fake Magritte with fishheads as seen through the eyes of Marjorie Nicolson. Howard pushes the limits of the poetic to include the cultural crossfire of the era that precedes our own, taking us down a slippery slope from the heights of literary history to a parody of the opportunisms of twentieth-century literary and art institutions.

In *Talking Cures*, two museum comedies continue this interrogation of the institutions that circumscribe and mediate the experience of art. "Joining" examines a rite of initiation into the professional atmosphere of museum administration itself, a rite that is exclusionarily gendered. A reminiscence in the voice of Dr. Gisela Richter, "the first woman to be made / (Associate) Curator of Antiquities," the poem describes her marginal position in the museum bureaucracy, literally and figuratively: "a desk was found down the hall / from the department" (TC 46). This position emphasizes that she "could never belong to the club, / not being one or any of the boys," but it also prompts an unexpected and empowering discovery when she finds a trunk labeled "ANTIQUITIES. MEMBERS ONLY." Punning on "members" as both in-crowd and anatomical appendage, Howard describes an array of sculptural fragments: "tray after tray of penises knocked off / how many missing statues, marble, flint / and granite [. . .]" (TC 46-7). Seeing these "knock-offs" enables Richter, and Howard's readers, to recognize the sham and bravura behind protocols of cultural authority.

Another comic send-up of museum policy and its contradictions, "Hanging the Artist" begins with the plural voice of the institution: "We just can't!" (TC 26). The poem presents the entreaties of a vexed curator who must preemptively censor a show by Yasumasa Morimura, a Japanese artist (b. 1951) known for his provocative self-portraits as Western pop culture icons. "Hanging the artist" indeed: the gallery becomes the artistic gallows as the curator cautions Morimura about the "powerful and possibly / traumatic impression these pictures of yours are apt / to make on our Houston art-lovers . . . [. . .]" (TC 26). They are museumgoers, the curator explains, who "tend to be repelled / by images that seem to question / or repudiate—you follow me?—the status quo / of gender" (TC 27). Like Julien Levy, this representative of the art establishment must mediate between a prudish public and a gender-bending iconoclast, but unlike Levy, the curator errs on the side of caution. Morimura is recalcitrant, forcing the curator to issue this ultimatum:

> the dress was up to her waist, the girl
>   was naked, I mean *you* were naked,
> and right in the middle of that big black bush of hair
>   was a prominent penis (I know you know
>   what *those* words mean). Morimura-*san*,
> believe me, the fact that it wasn't a *real* penis
>   makes no difference whatever. The Houston
>   Contemporary Art Museum
> will not show Marilyn Monroe with a penis, now
>   Get. That. Straight. [. . .] (TC 28–9)

The poem humorously exposes two problems with museums—the numbing of the sensual, the conservatism of cultural recognition—by upending the terms of cultural value and mocking institutional efforts to control those values. The curator may have the last word, but the unruly energies of Morimura's work have already unsettled the decorous atmosphere of cultural uplift that "art-lovers" expect.

Moments of museum comedy like these continue to appear in Howard's newest work. In "In Loco Parentis, 1963," which was published in the Winter 2003 issue of *The Georgia Review*, Blanche Knopf remembers crossing paths with Joseph Conrad at the Louvre in 1923, under the Victory of Samothrace. Throughout Howard's rich and varied body of work, museum scenes signal nodes of tension between the "post-" and the "pre-," ambivalent but entertaining interludes that invoke the cultural forms of the past and wryly interrogate their relations to the present. In "Lining Up," Howard reduces the apotheosis of Rodin to the contradictions of contemporary life in the space outside a museum's ticket booth. In "Even in Paris," he enlists Monet and Whitman to dazzle Wallace Stevens in the Louvre, only to leave him in a comic stupor. In "Family Values V," he skips a critical generation and locates his interrogation of the institutional conditions for aesthetic value in the cultural terrain of 1947. In museum moments, Howard casts back for an earlier example or type to expose the mechanisms and mediations inherent in contemporary aesthetic experiences. If it is typical, as Christopher Beach has argued, invoking Bourdieu, that a writer who embraces "literary" and "artistic" values tries to conceal "any consideration for the means by which a work is produced, the institutional authority that allows its production, or the role of the 'cultural businessman' [. . .] in promoting dissemination or consecrating that work" (198), Howard's work provides a postmodern counterexample. Howard's poems are *about* this very process of institutional authorization and consecration, and "cultural businessmen" (and women) are among his characters. Because he is so witty and extravagant as he goes about it, we tend to forget that Howard's mode is one of dissent. He stages comedies in the space of the museum to disrupt ritual postures toward art.

Museum encounters in Koch's and Howard's work reflect both irreverence and delight. Koch summarizes this fundamental exuberance in one of his "poetry comic strips." "Going to the Museum Comics" presents a gallery layout with boxes representing large and small artworks, including one "REALLY BIG PICTURE." The captioned responses to all of these works, in different sizes, are "HAHA!" and "HAHAHAHAHA!" (*The Art of the Possible*, 20–1). In my discussion of "museum comedies" in this chapter, I have stressed that laughter in museum space coincides with keen insight into art and its institutions: for both Koch and Howard, the museum is a lens that focuses their critical responses to the institutional framings of aesthetic experience. Recognizing the importance of the museum setting in their work highlights their interrogative stances and allows us to read their work against the grain. The dual canon in contemporary poetry aligns Koch with avant-garde critique and Howard with literary homage, but their museum comedies demonstrate that both poets explore the zone between these poetic positionings. At times, on these museum thresholds, their poetic paths seem to intersect and then move in opposite directions, departing from the trajectories that seem characteristic of their work. We find that Koch, despite his oppositional posture, can be a more romantic museumgoer than his avant-garde sympathies might lead us to believe. We find that Howard, despite his literary seriousness, is a more ambivalent museumgoer than his artful displays of high-cultural themes suggest. The museum comedies of these two poets offer a forceful corrective to readings of their work that would place them on the polar extremes of the contemporary poetic spectrum, reminding us that Koch and Howard, over the course of their long careers and from multiple vantage points, engaged and challenged the oppositions of their moment with comic aplomb.

Chapter Four

"Someone (A Woman) Watching": Site-Specific Ekphrasis by Three Feminist Innovators

Amidst the proliferation of poems about the visual arts in recent years, the ekphrastic work of three poets commands attention for its inventive form, sensory intensity, and critical acuity. Cole Swensen, Kathleen Fraser, and Anne Carson have expanded the possibilities of the genre by approaching the visual arts through "site-specific" practices: they compose ekphrastic poems through strenuous engagements with the place of the encounter. In this chapter, I adapt the concept of site specificity from its usage in the visual arts to emphasize the ways that these poets, as they focus their attention on art objects, produce work "that defines itself in its precise interaction with the particular place for which it was conceived" (Fineberg 322). To borrow another definition from art criticism, these poets strive to address and "incorporat[e] the physical conditions of a particular location as integral to the production, presentation, and reception of art" (Kwon 1). In the three examples I present, the particular location in question is a museum, a site that invites a complex series of meditations on art objects and their contexts in the past and present. In "Trilogy," from *Try* (1999), Swensen scrutinizes the acts of attention through which she interprets the *noli me tangere* theme in a museum gallery and its urban environs. In "Giotto : ARENA," from *when new time folds up* (1993), Fraser reads Giotto's frescoes at the Arena Chapel in Padua through the layers of commentary that surround them. In "Canicula di Anna," from *Plainwater* (1995), Carson enters the fortress that marks the medieval center of Perugia to dramatize an encounter with Il Perugino and his muse.

Applying the term "site-specific" to a literary context, I underscore these poets' attentiveness to the material details of the ekphrastic setting, and

their foregrounding of necessarily local perspectival shifts, even as I acknowledge that the most strictly literal sense of this term does not pertain. Poems
are made of ink and paper, not steel or stone in a landscape, and are portable
and reproducible in ways that site-specific works of art, such as Richard
Serra's *Tilted Arc* (1981),[1] are not. Nonetheless, I do not expand the term to
encompass all "discursive formations" that relate to place. In her genealogy of
site specificity, Miwon Kwon chronicles this expanded usage: over the past
30 years, "the operative definition of the site has been transformed from a
physical location—grounded, fixed, actual—to a discursive vector—
ungrounded, fluid, virtual" (29–30).[2] Once it becomes this broad, however,
the term loses force—what work of visual art, or poem for that matter, is not
a more or less grounded "discursive formation" (30)? In order to highlight
the particularity and significance of these poets' innovations, I use the term
to signify a "material formation" and not a vague conceptual notion. When
Althusser, whom I will invoke again with respect to Swensen, makes a case
for the "material" existence of ideology, he must acknowledge that it "does
not have the same modality as the material existence of a paving-stone or a
rifle" (156). I must make a similar concession in claiming that some poems
are "site-specific," and I draw on Althusser's logic to argue for the term's usefulness in this context: "I shall say that 'matter is discussed in many senses,'
or rather it exists in different modalities, all rooted in the last instance in
'physical' matter" (156). My use of "site-specific" relies on this ultimate root-
edness in physical matter, and is intended to reflect the term's earliest phe-
nomenological or experiential meaning at one verbal remove: these poems
record sensory, emotive, and linguistic participation in a physical location.
They are site-specific, and compelling, for their emphasis on the media of
aesthetic experience, for their foregrounding of the institutional conditions
that frame and illuminate those artistic materials, literally and ideologically,
and for their vigorous historicity.

    For Swensen, Fraser, and Carson, site specificity is also gender speci-
ficity. These poets are attuned to the role of a perceiver whose gender is often
announced, a perspective Swensen identifies in one poem, characteristically,
as "someone (a woman) watching" (*Such Rich Hour* 21). Reading their work
requires us to notice the gendered identity of the perceiver, but it also
requires us to acknowledge that the role gender plays in these poems is not
straightforward: the information given in Swensen's parenthesis may be sup-
plementary, or secondary, or emphatic. The "someone" who is watching is "a
woman," but how and why that descriptor matters is not self-evident. It has
become a commonplace in visual-verbal studies to claim that gender is central to the ekphrastic exchange, that ekphrasis unfolds along the lines of a

"commonly gendered antagonism" (Heffernan 7). In this chapter, I turn to poems that require us to reconsider and complicate that commonplace.

Two quotations will suffice to present its familiar terms. James Heffernan begins his literary history *Museum of Words: The Poetics of Ekphrasis from Homer to Ashbery* (1993) with this explanation:

> the contest [that ekphrasis] stages is often powerfully gendered: the expression of a duel between male and female gazes, the voice of male speech striving to control a female image that is both alluring and threatening, of male narrative striving to overcome the fixating impact of beauty poised in space. (1)

W. J. T. Mitchell, in "Ekphrasis and the Other" (1994), offers a similar paradigm in more theoretical language. Ekphrasis involves a

> suturing of dominant gender stereotypes into the semiotic structure of the imagetext, the image identified as feminine, the speaking/seeing subject of the text identified as masculine. All this would look quite different, of course, if my emphasis had been on ekphrastic poetry by women. (181)

Both Heffernan and Mitchell extrapolate this pattern of "male-looks-at-female specularity" (DuPlessis 81) from canonical examples—Keats's "Ode on a Grecian Urn," Shelley's "On the Medusa of Leonardo Da Vinci in the Florentine Gallery," Browning's "My Last Duchess"—and both read ekphrastic texts as descriptive and expressive of a clear distinction between two gendered parties, whether those parties are in opposition (in Heffernan, "the expression of a duel") or in tense fusion (in Mitchell, a "suturing" signified by the compound "imagetext"). Both argue that interart encounters often follow and reproduce a hierarchical structure of dominance—male over female, text over image—and both align "male" with text, speaking, and beholding subject, and "female" with image, silence, and art object. The rigidity of this scheme is perhaps what provokes Mitchell, with a dismissive "of course," to assert that women write ekphrasis differently, but he does not explain why this might be so. He goes on to state that "the difference, I would want to insist, would not be simply readable as function of the author's gender" (181), but he offers no additional measure by which to read that difference, nor a mechanism through which women writers might intervene in the dominant gendered paradigm. To posit the importance of gender to ekphrasis, both Mitchell and Heffernan rely on essentialized distinctions between male gazes and feminized objects.[3]

As an alternative, to correct the reductive terms of gender difference and gendered dominance in ekphrastic studies,[4] I propose that we read these late twentieth-century ekphrastic poems by women as less descriptive than deictic. The verbal strategy that enables their site specificity and gender specificity, I will show, is a pervasive and self-conscious amplification of *deixis,* a foregrounding of the "here and now" of their utterances through verbal markers of the present time, location, and participant role of the speaker.[5] Émile Benveniste explains that together with the personal pronouns,

> the indicators of *deixis,* the demonstratives, adverbs, and adjectives [. . .] organize the spatial and temporal relationships around the "subject" taken as referent: "this, here, now," and their numerous correlatives, "that, yesterday, last year, tomorrow," etc. They have in common the feature of being defined only with respect to the instances of discourse in which they occur, that is, in dependence upon the *I* which is proclaimed in the discourse. (226)

Deixis is the equivalent, in a verbal medium, of the materials and tools that tether a site-specific visual or sculptural work to its landscape, street, or room: "*here* and *now* delimit the spatial and temporal instance coextensive and contemporary with the present instance of discourse containing *I*" (Benveniste 219). Benveniste distinguishes deixis from what he calls the "historical" tense, the mode of objective description and exposition that arises from the impersonality of third-person narration (Bryson 2).[6] Unlike deixis, which operates along an "I-you" axis, the historical tense is "the mode of utterance that excludes every 'autobiographical' linguistic form": "Events that took place at a certain moment of time are presented without any intervention of the speaker in the narration" (Benveniste 207, 206, qtd. in Bryson 2). Mitchell and Heffernan tend to read ekphrasis as if it were always written in this objective mode; that is, they read ekphrasis as expressing or narrating a gendered conflict that occurs between two parties whom we can label, through analysis or observation, as "he" and "she."

The poets I present in this chapter write ekphrasis from the opposite linguistic tactic. By making frequent and prominent use of deixis—Swensen even calls attention to this preference in a poem titled "Deictic," part of a series after Hieronymus Bosch—they highlight, interrogate, rupture, and reassemble the interventions of the speaking and seeing "I" in relation to its particular place and time. In making this claim about their deictic strategies, I do not suggest that other poets—Keats and Browning among them—do not point to "this" and "here" as they situate their ekphrastic speakers (positionality is indeed a

classic ekphrastic preoccupation), but that there is a particularly high con-
centration of importantly placed deictic markers in these three poets' work.
Noticing these deictic markers enables us to understand more fully the gendered underpinnings of ekphrasis generally, and postmodern feminist
ekphrasis in particular. Nor do I claim that these poets use the mode exclusively—they narrate stories of gendered figures, use the third person, and
offer vivid descriptions, to be sure, but they cross these stories and images
with interruptive, extraneous, and often disorienting reminders of the "here
and now" of their speaking and seeing. What is significant about Swensen's,
Fraser's, and Carson's ekphrastic work is not the fact that they are women,
but that they themselves interrogate gendered tensions in the particular deictic sitings of their visual-verbal exchanges. Mitchell, I would argue, is asserting the wrong kind of difference: the critiques these women poets offer are
not deployed through their own gender identities, but through linguistic
innovation. Their inventive uses of deixis enable them to exercise an interrogative relation to the roles of gendered subjects within a museum context,
and to reframe ekphrasis along lines of vision that are alternately focused and
diffused.

"I DO HERE SUSPEND THE HERE AND THIS":
COLE SWENSEN'S VISION OF CONTACT

In "To Writewithize" (2001), an essay for *American Letters and Commentary*,
Swensen proposes new interactions between poetry and visual art, issuing a
challenge to write "with eyes," as in "to hybridize" or "to ionize" (122).[7] The
essay offers a summary of Swensen's poetics, underlining the importance of
visual media to the creation of charged lyrics where "eyes," doubling as the "I"s
of a volatile subjectivity, multiply and recombine. Throughout her work, writing "with eyes" entails an ekphrastic mode, a method announced in the opening
line of her early book *New Math* (1988): "Let me get this picture straight" (13).
Beginning with deictic markers—the first-person pronoun ("me") and demonstrative adjective ("this")—Swensen establishes the speaker's perspective relative
to a picture that emerges through glimpses of color, facial expression, and
atmosphere. A later poem titled "Deictic" declares the centrality of this grammatical strategy to her ekphrastic poetics. It follows two poems after Bosch
titled "Here," and it is playfully cryptic about the instability of attempts to pin
down visual referents, especially when the personal pronoun is interposed:

> So you can be here now and be it at the same time and you can take it
> with you and when you're gone, so will be my eyes replaced by eyes. I

smile with a certain amnesia, an amnesia that knows itself certainly, certainly they all lived here once and they were beautiful then but only then. (*Try* 29)

In a Steinian welter of pronouns, the "you" who is "here now" becomes both an "it" and an "I," both object and subject. Understanding the painting requires not only recognition of "it," the art object and its historicity ("they all lived here once"), but also recognition of the "I" who sees, smiles, forgets, and "knows itself." Swensen explores the reciprocity of "I" and "you" in her mode of address as she examines the object of beauty before her.

Emphasis on deixis enables Swensen simultaneously to dwell within and to examine her acts of attention vis-à-vis visual works. It enables her to "write-withize"—to write in an ekphrastic mode that produces "ionized" works by manipulating and "suspend[ing] the here and this" (*Such Rich Hour* 43). In her notational stops and starts, fragmentary accretions and disconnections, Swensen "sieves" ekphrastic observations (the verb is hers) through awareness of their relational complexity, subjective translation, and linguistic construction. The point of view that emerges is one of "someone (a woman) watching" (21), a perspective attuned not only to the object or scene being watched, but also to the stance of an observer whose gender is bracketed but apparent. In this section, I show how Swensen negotiates the tensions of this gendered perspective through a deictic ekphrastic mode that positions the observer within a web of glances, readings, rebukes, and longings. The site specificity of this mode is exemplified in "Trilogy," where a physical confrontation with a work of art in a museum dramatizes the interpellation of the speaker as a gendered subject and unsettles the terms of ekphrastic looking.

To begin, a trio of moments from *Such Rich Hour* (2001) illustrates the importance of deixis to her ekphrastic method.[8] The book is an extended meditation, in a calendrical sequence, on female experience as represented in the *Très Riches Heures du Duc de Berry*, and on sources that describe the work's historical and cultural context. Throughout the sequence, in which the fifteenth-century book of hours inspires an ekphrastic poem dated the first of each month, the process of looking leaves traces of the looker, as in this description of the illustration for the month of March:

> [. . .] A small dog is running up to the road
> in front of a flock of sheep in front of a large man carrying
> something (we can't see what) (sheaf?) (shearing?) holding (soft)
> against himself. (29)

The passage begins as a third-person objective account in which elements in the picture are consistent in relation to one another, but it soon veers out of the picture plane with a first-person admission of uncertainty. "We" cannot ascertain the significance of a mark on the visual surface, and this uncertainty provokes speculation at the level of the *verbal* surface: "sheaf" leads to "shearing" via the words' sonic values as much as via semantics. We are in the midst of a process of testing out words for images, the run of "ing" words (running, carrying, something, shearing, holding) propelling the lines through the cognitive glitches. Drawn into the description in this way, we are pointed to a deictic center, a reference point that emphasizes the participant role of the beholder-speaker and reader.

In the ekphrasis of April's betrothal scene, reliance on deictic reference enables Swensen to stress the beholder's participation in an examination of the gender relationships that the work of art depicts. Beginning with "this" (here an example of "empathetic deixis," an indication of emotional proximity to a referent), she considers the book's representation of a patriarchal practice: "This is how they lived: the dialogue was staged / and every woman given thus thus said / Pardon me" (42). Exposition again veers into the first person, a "me" who could be either the speaker or a party in the nuptial dialogue. Addressing a historically gendered exchange—"every woman given thus"—Swensen then reads the picture for its symbolic significance:

> in a field of green (read: early love) (ground from the malachite of
> Hungary) wearing blue (both) (doth bet Fidelity against Eternity) made of
> lapis lazuli brought
>   all the way back from the Black Sea [. . .]. (42)

Because they are the backdrop and costumes for a social and religious ritual, the green landscape and blue garments take on the meanings "early love" and "fidelity." The imperative "read" and the series of parentheses direct our attention to the speaker's participation in the process of interpreting these meanings, offering further information about the geographical sources of the artistic media, and the ceremonial language that might accompany the scene.[9] In this way, Swensen's ekphrases never get far without reminding us of the observer—the writer—who culls and interprets such details, understanding their materiality and historicity only through the lens of her own perspective. She is ever attentive to the relational nature of these acts of looking and describing, as in this evocation of the May scene:

He turned
        toward a castle just visible beyond the trees,
and pointed and smiled, but I couldn't hear what he said. (50)

In this humorous shift, narration of the imagined actions of a figure in
the picture returns to the first-person observer, momentarily dissolving
the distance between beholder and painting. Swensen invokes the traditional ekphrastic aim of envoicing a mute painting, only to deny that
possibility in a gesture that grounds the description in a deictic moment
of perceiving.

These three moments represent efforts to write ekphrasis in a new way
by involving and exhibiting the first-person perceptual mechanisms that
operate in approaches to art. Swensen explains that traditional ekphrasis

> accentuates the separation between the writer and the object of art. The
> writer not only remains figuratively outside the visual piece, but often
> physically in opposition to it, i.e. standing opposite it, in a kind of face-
> off, in a gallery or museum. And often the physical stance echoes the
> mental: despite the apparent homage, there's frequently an element of
> opposition, a tinge of rivalry and/or challenge inherent in this mirror-
> ing—can the poem match the painting in impact? And/or be a "faith-
> ful" translation of it? And there's a tinge of protection—writing is used
> to keep art at a safe distance, to keep it sealed in its frame, demonstrably
> the "other" of poetry. ("To Writewithize" 122–3)

Challenging this traditional stance—the bipolar paradigm in which the
beholder addresses the art object as other—Swensen proposes instead that
writers approach art with greater awareness of their complicity in the process
of exchange. She prefers

> works that don't *look at* art so much as *live with it*. The principal differ-
> ence here is not in the verb, but in the preposition. There's a side-by-
> side, a walking-along-with, at their basis. (123)

Praising the work of Susan Howe, Johanna Drucker, Carol Snow, Laura
Moriarty, Mei-Mei Berssenbrugge and others, Swensen stresses cohabitation
rather than confrontation, contiguity rather than opposition, peripatetic
movement rather than contest. In her own poetry, the effort to "live with" a
work of art entails locating the experience of it in the work's own history and
in the sensations and institutions of the speaker's moment. The process leads

her to re-examine the place where that "walking-along-with" often tran-spires—the museum.

In "Trilogy," the second sequence in her book-length ekphrastic proj-ect *Try* (1999), Swensen exposes the gendered tensions of a museum "face off" and recasts the encounter as an unsettling scene of coming to subjectiv-ity. *Try* addresses a wide range of artists, including Giotto, Orcagna, Ben-venuto, Mantegna, Bellini, Tintoretto, Bosch, Memling, Rodin, and Olivier Debré, in poems composed of short-lined stanzas interlinked with prose passages, each sequence structured in groupings of three. This form suggests verbal collages "arranged in triptychs" (Conoley 1), triple presentations of scenes and stories that preserve the outlines of their painterly antecedents despite their fragmentary composition. "Trilogy" is divided into "One," "Two," and "Three," with each of those parts subdivided into three sections headed with Arabic numerals. In these symmetrically ordered sections, the poem presents this three-part plot in fragments: a woman goes to a museum to study a painting and breaks a rule; the woman leaves the museum and contemplates her solitude in relation to an urban environment; the woman returns to the museum with a heightened self-consciousness through which intimate and objective perspectives coalesce. Swensen offers some instruc-tion as to how to read this story in fragments in the opening pages of *Try:* "viewer hold me to you, lace this fracture to a future" (3). This book's bal-anced triptychs hinge on an "I-you" axis, with an undecidable "me" implor-ing an equally indeterminate "you" to maintain the close proximity of an embrace (a troubling proximity, as we will see). Only then, the poems sug-gest, can the "fractures" of their verbal disjunctiveness yield a "future" of communication to the reader. As Gillian Conoley puts it, Swensen attempts to capture "our speed of sight, the filmic surface of contemporary vision, our ability to pan, scan, splice, fragment, and make whole again" (3). Pre-senting this spliced and segmented vision, the collages suggest a unifying if not unified point of view—a traveler, a museumgoer, not quite at home in her surroundings. She, like the speaker of many of Elizabeth Bishop's poems, looks at unfamiliar things from multiple perspectives, penetratingly but not absolutely. (In "Poem," Bishop substitutes "looks" for "visions," preferring its more tentative and provisional, less "serious" connotation [*Poems* 177]).

I say "she" deliberately. Reference to the gender of the speaker is fluid but ubiquitous in *Try,* from the generalized perspective of "Every woman's eye" (4) to the particularity of a "Person, pale, female" (21). "Trilogy" opens with a "she" who comes face to face with a male figure of authority—a museum guard:

1

The guard peers closely
at the painting. Count.

The fingers. The figures. The
strange sweep from waist

to chest to head. His hand reaches out
within a second of

2

She sweeps upward. Up
to where the gold sky might (13)

From the moment of this first spare sketch, the ekphrastic encounter is trian-
gulated rather than polarized: the presence of the guard shadows the
speaker's scrutiny of the canvas. As the museumgoer studies a figure of a
woman, examining the "strange sweep" of her body "from waist / to chest to
head," the imperative to "count" the painting's elements doubles as the
poetic task of arranging these observations in predominantly four-word
lines. But this process of objective enumeration is interrupted when the
speaker breaks the first rule of museum protocol, as the next section
reveals—she touches the painting. The noun "sweep" becomes a verb
describing the museumgoer's action: the guard instantly interferes, and "She
sweeps upward. Up / to where the gold sky might." Both parts of the result-
ing action—the guard's hand reaching out, and the museumgoer straighten-
ing up to look at the "gold sky" of the painting—are left as incomplete
sentences, suggesting that the museumgoer's cognitive as well as physical
advances are abruptly curtailed.

     The museum guard's prohibition of touch catalyzes a meditation on a
famous instance when a rule of "You must / not touch" (49) applies: "What
would the touch / if it did not first / / run up against / a man who is in the
end a man" (13). This tautological gendering, in which the speaker has "run
up against" male opposition, alludes to a scriptural account that is contem-
plated throughout *Try*—Jesus's rebuff of Mary Magdalene when she recognizes him on the morning of the resurrection. "Triad," the poem
immediately preceding "Trilogy," opens with a description of "Noli Me Tangere, *Unknown Spanish Painter, early fourteenth century*" (7). Another poem

in the book, titled "Noli Me Tangere," is an extended meditation on this staple of Renaissance and Baroque painting, referring to works by Bronzino, Poussin, Rembrandt, and others. Exploring artistic depictions of this scene, Swensen ponders the hurtfulness of Jesus's statement ("Mary just turns her back on him and cries" [48]), but she also reconsiders its context. In the Gospel of John, Jesus's point is that the man is *not* in the end a man: "Do not cling to me, for I have not yet ascended to the Father" (John 20:17, Suggs 1391). Mary has tried to touch a "man" who occupies a liminal position between incarnated God-as-man and glorified divinity. Swensen succinctly notes the paradox: "If you profane this is not flesh" (8). Mary's gesture is both a profanation, because the human touch (of a reformed prostitute) would defile the sacred, and an impossibility, since Jesus is inexplicably no longer "flesh" that can be embraced, despite his material appearance. Meditating on the paradox in this sequence of poems, Swensen also makes a crucial observation: "So rarely in these paintings is anyone speaking" (41). She is not making a naive point about painting's muteness, but observing that the bodies and gestures on these canvases do not signify conversation or outcry. This omission, she suggests, is strange, since the biblical passage continues with Jesus's immediate qualification of his rebuke: don't touch me, he says, "But go to my brothers, and tell them that I am ascending to my Father [. . .]" (John 20:17). He dispatches Mary to *tell*. As Swensen describes various paintings that depict this juncture, she examines tensions between touching and telling—between the desire for physical contact, intimacy, and immediacy, on the one hand, and the imperative to inform, narrate, and reveal, an imperative that enforces a hands-off distance, on the other.

In "Trilogy," Swensen literalizes this tension in microcosm in the museum scenario in which she sees these works of art. The speaker defies the museum rule against touching, tells about the transgression, and infuses it metonymically with the emotive intensity of the biblical account:

3

She touched the painting
as soon as the guard

turned his back. Respond.
I said turn around. I

screamed, I drowned, I
thought you were home.

I touched the surface of the canvas.
It was I, the sound of salt. And fell
and is still falling through a silent earth. (14)

The section begins with an effort to relate the episode in third-person past-tense narration, but a command to "respond" insinuates itself, compelling a shift into a deictic mode. The first-person pronoun—repeated here six times in five lines—signals participation, desperation, and confusion. The deictic "I-you" axis of these lines suggests an unstable trajectory from museumgoing speaker to the figures in the paintings she sees, causing the ordinary expression "I thought you were home" to carry some of Mary's astonishment at finding Jesus both "home"—restored to life—and completely alien. This seeing self is full of longing in excess of the museum situation, and the resultant "fall," as the present perfect tense of "is still falling" indicates, is dire and encompassing. Yet we remain in the museum setting, where the first-person speaker admits, "I touched the surface of the canvas."

The moment of transgression in the museum is highly charged, loaded with the biblical freight of "falling" and the sensuality of the synaesthetic image of "the sound of salt" in apposition to the "I." We are in the presence not only of Jesus and Mary, but Lot's wife: fleeing a scene of punishment, she defies the command not to turn back toward Sodom: "But Lot's wife looked back, and she turned into a pillar of salt" (Genesis 19:26, Suggs 28). Swensen's speaker, momentarily disturbing the usual museum order of things, aligns herself with this figure who is paralyzed as a consequence of a hungry gaze and a refusal to obey. She too is *arrested* by her gesture, stopped short by the security guard: "I said turn around." The exchange replicates what Althusser calls a "*mise en scène* of interpellation" (165), a moment in which the intervening call of institutional authority (a police officer in Althusser's account) functions as "the hailing or interpellation of individuals as subjects" in ideology (163). This museum scene, in which "I"s proliferate frantically, suggests that the speaker becomes cognizant of herself as a disciplined subject and attempts to adjudicate the physical and linguistic circumstances of that becoming: "the hailed individual will turn round. By this mere one-hundred-and-eighty-degree physical conversion, he becomes a *subject*" (163). Stressing that this subject is a *she*—a female subject aligned with Mary Magdalene and Lot's wife—Swensen choreographs a transgressive ekphrastic moment that refuses "to keep art at a safe distance," but that cannot be sustained. The speaker attempts to close the gap between observer and object, the guard intervenes to reinstate and enforce that gap, and the speaker "turns around" into an unsettling recognition of her incontrovertible status as a gendered subject.

This attempt to come into contact with art, to "live with it," has failed—the call cannot be resisted, and the subject must and does stand back from the painting—but from this moment onward, the speaker casts a backward glance at the seeing self each time she looks outward. The second movement of the poem carries this self-consciousness out of the museum and into the circulating social space of the public square. Passing a café, the speaker observes: "Look, he's speaking, leaning over to his neighbor" (15). The poem continues to map the vectors of the speaker's gazes, gazes directed not at mute or feminized objects, but at men and women speaking and interacting. The next gaze is turned back on the speaker as she situates herself in this site: "She crosses the square in a bright red coat. / Look how they look at her, look up / / from their talking [. . .]" (15). Swensen uses the word "look" in three different ways in this line—as an imperative directing attention, as looking "up" in interruption, and as looking "at" an object of desire. The act of looking is never simply bipolar: her position shifts from observer to observed, and she becomes aware not only of her own looking (as a woman), but of being looked at (as a woman). The mood of the scene, carrying over from the last, is one of unnerving isolation and disconnection: "There are people crossing the square / arm in arm, in threes and fours and alone in great numbers" (15). Enumerating elements of the street scene as she did elements of the unnamed painting, Swensen shifts from generalized observation ("there are people") to admission of homesickness: "She crosses the square on her way home. / She will not stop at a café, she will not talk, she will just go home" (16). The parallel sentences and repeated pronoun "she" suggest that as the speaker navigates an urban social space, she finds herself caught in the uncomfortable awareness of being both gazer and gazed-upon.

As "Deictic" puts it, "you can be here now and be it at the same time" (*Try* 29). The speaker identifies herself as a subject in a particular place and time (the deictic center "here now"), but also recognizes herself as an "it," one object among many in this place: "And one by one, they looked up and watched" (18). Similarly, in the earlier museum passage, this simultaneous occupation of both agency and objecthood is encapsulated in reverse in the clause "It was I." The "it" becomes the self, the "I" who guiltily assumes agency—"I touched the surface of the canvas." Swensen's speaker, like Althusser's friend at the door in his scenario of ideological recognition (161), automatically responds to a question of "Who's there?" with the answer "It's me." Once the speaker has been interpellated as a subject, called to as a "you," she fills the space of a "she," inhabits the subjectivity of a woman who can be reprimanded in a museum and watched while she crosses a square. Throughout her poetry, Swensen attempts to keep that "you" mobile, using it variously to represent an intimate other, a lyric addressee, an indeterminate persona, and a manner of talking-to-oneself, but she cannot escape a "you" of chilling particularity—you there, stop what you're doing. Even as the pronouns shift relentlessly, the "you" of "Trilogy" recognizes herself as a gendered subject isolated by that defining call—a subject who must keep her hands off paintings and her eyes looking homeward.

The final and most sensorily evocative section of the poem returns to the museum in the wake of this acute self-siting. In consecutive couplets, the speaker scrutinizes a painting and recalls another ideological configuration that the street scene has presented, the contrasting site of "home":

1

The minute progressions between grey and black
becoming one against the red that stares back.

Because she knows they are watching, she will not turn around.
Home is a sound repeated to solid, to something that will hold.

Look, there goes a man with his left hand left lightly
on the head of his child. There they go. (17)

The section begins by noting the painting's contrasting colors, with the rhyme of "black" and "back" underscoring the merging of tints in proximity to one another. A brief mention in the previous section of Josef Albers's *Interaction of Color* (1975) corroborates this emphasis on color relativity, suggesting a relational as opposed to fixed visual space.[10] Swensen then cuts quickly from visual analysis to the relationality of her subject position: the gazes of strangers make the speaker long for familiarity, for a site that unifies rather than estranges. The statement "Home is a sound repeated to solid" is ventured as an aphorism, suggesting that repetitive insistence on "home" can transform the word ("sound") into thing ("solid"), can bring the concept into material existence. The proposition finds its objective correlative in the image of father and child, the symmetry of "left hand left" and the custodial "his" suggesting the tender but possessive enclosure of (patriarchal) familial ties. The image contrasts the isolation engendered by the disciplinary moment in the museum, but it is only a deictic passing glimpse: "Look, there goes" and "there they go." It is the only time in the poem when hands are permitted to touch, but it too is a closed circuit, a vision of "home" in which the terms of participation are static and in which the solitary speaker cannot make contact: "She was one" (18).

In the context of this longing for and distance from "home," the ekphrasis that follows is imbued with yearning:

> In the painting, all the reaching hands are growing.
> In the gallery, everything was green and gold and red
>
> Made the sun, though deep, cut through:
> Within the door was a window; within the window, a jar.
>
> Inside the jar, carefully there, the love need not be
> assigned in order to fix and ignite. (17)

In this postmodernist reworking of Stevens's "Anecdote of the Jar," two prepositional phrases announce the site specificity of the ekphrasis: we are looking at a painting in a gallery, at an artwork and environment that color each other. The painting's "reaching hands" (possibly Mary Magdalene's) seem to be extending into gallery space, and its "green and gold and red" hues affect the natural lighting that illuminates it—it "made the sun, though deep, cut through." The couplet form, in which alliteration further unifies the lines (gallery / green / gold; within / window / within / window), suggests quantities of focused attention made possible by this confluence of illumination from artwork and site. This intense focus yields the telescoped image of the jar within the double frame of the window and door: a view through a museum passageway converges on an image that replaces Stevensian aesthetic dominion with a gestational interior. The speaker then asserts the presence of "love" in that enclosure—located "carefully *there*," in deixis—as if the lens of concentrated looking had garnered the intensity of the multiple gazes that cross this poem, and enabled them to "fix and ignite" without the poles of subject and object: "the love need not be / assigned [. . .]."

This image of "unassigned" love within multiple visual frames marks the poem's climax, and what follows in the conclusion is a denouement and rewriting of the earlier interpellative scene. Swensen has the *guard* turn around and the museumgoer speak the interdiction:

When the guard turned around, the gesture was gone.
A woman stood back and said no.

She stood back, looking at the painting and said isn't it fine
that a woman wearing red could arrive at a gold sky. Remind.
Or else in falling. And nothing broke. The rift
shifts open the devout. A finger that exceeded number, a
fingertip. (18)

Defusing the earlier passage by suggesting that the guard never saw her transgressive "gesture," the museumgoer presents herself as conforming to museum protocol and politesse: "She stood back, looking at the painting and said isn't it fine [. . .]." She rewrites the story as if she has gotten away with her touch, made contact but gone undetected, and then resumed proper behavior under surveillance. Officially forbidden to touch, she accepts that she can only *tell*, and makes an appropriate observation about the aesthetic effect of the painting's contrast of red and gold ("isn't it fine"). But at this point in the poem the observation takes on a peculiar resonance—the speaker herself has been identified as a "woman wearing red." The word "remind" calls a halt to this conciliatory conclusion and draws the poem back through its preoccupations with "falling," with the "rift" between Jesus and Mary, and with the "fingertip" that touches what it shouldn't. What has happened, as acts of looking and telling have crossed and re-crossed this poem, is that the image of the woman in the painting has blurred with the image of the speaker. Both are identified by red garments and visualized against a "gold sky," both are forbidden to touch, and both are positioned as simultaneously observing and being observed. Shifting among pronominal and demonstrative markers, Swensen's deictic ekphrastic mode has momentarily allowed the art object to slip into the beholding subject and back out, into museum space.

"A woman stood back and said no" (18): at the heart of this poem, and Swensen's work more generally, is a gendered gesture of defiance, a refusal and reconsideration. At the level of "Trilogy"'s plot, a solitary woman's defiance takes the form of stares and glances alternately projected outward with longing for contact, shot back at the intruding eyes of guards and café-goers, and turned inward toward the self. At the level of ekphrastic inquiry, with which Swensen is engaged throughout *Try*, this defiance takes the form of a re-envisioning of the interrelationships between looking, touching, and telling, a re-envisioning that accounts for the intersecting trajectories of those actions. In light of the exposed tensions of a scene of interpellation, this defiance entails not an attempt to escape from ideology but an effort to examine the responses and gestures with which a gendered subject takes her uncomfortable place within it. Throughout Swensen's poems, deictic markers are reference points that highlight the participant role of the beholder-speaker in the scenes she sees, and they complicate any attempt to align her acts of attention with bipolar terms of gazing subject and mute object. Instead, Swensen's arrays of fragments suspend us in the surface tension of the utterances that are constitutive of subjectivity. Her volatile pronouns, inextricable from their contexts and their particular sites of interaction—the various "he"s and "she"s that come into contact with "I"s and "you"s—destabilize the stance of "someone watching." That someone is often "(a woman)," and she watches with an avid, distractible, responsive gaze.

READING ARENA:
KATHLEEN FRASER'S PALIMPSEST DIALOGUE

In "Giotto : ARENA," ekphrasis becomes a process of probing through texts as well as scrutinizing visual works. Through intricate citation, Fraser builds the poem within and against the contours of an existing literary and art-historical terrain: Dante, Vasari, and Ruskin have been there before. Fraser's note about the poem's composition emphasizes the site specificity of this process:

> Reading critical assessments of a "master," after being in the actual presence of Giotto's work; understanding from Ruskin and Vasari that Giotto's ongoing friendship with Dante involved painting the very same neighbors into Arena Chapel frescoes as those being written into Dante's *Inferno*. Wanting to foreground the meanings inherent in "faulty copying" (as typos visualize them); playing error against mastery—the fixed Byzantine model of 'perfection' as tripped up by Giotto's break from type. *(il cuore* 195–6)

Stressing the importance of "being in the actual presence of Giotto's work," Fraser describes an encounter with the work of a specific artist in a specific place—Giotto's frescoes in the Arena Chapel in Padua, built by Enrico Scrovegni in 1303 on the site of a Roman arena. She studies these works of art in their original location, an approach that critics of the museum going back to Quatremère de Quincy would applaud as an appropriate return to context, but her site visit is also an encounter with the layers of interpretation and commentary that have since enshrined those works as masterpieces. The Arena Chapel in 1990 is itself a museum, of course, a cultural site designated as such through inherited appreciations, and the tensions engendered by this setting provoke Fraser to sift through the visual and textual fragments she finds there.[11]

As Fraser documents the experience of viewing Giotto's work in this place, a critical inquiry unfolds about tensions between "mastery" and "error"—between notions of ideal form and artistic perfection, on the one hand, and the destabilizing presence of the actual and the particular, on the other. Fraser dwells on the latter—the mistakes and momentariness, impatience and idiosyncrasies, of the perceiving mind and body. Against the "mastery" described in the art-historical narrative of Giotto as legendary master artist in this legendary place, a narrative upon which the museum known as the Arena Chapel depends, Fraser offers the here and now of seeing, reading, copying, and writing. The "errors" inherent in performing these actions "in person" are announced on the opening page of the poem: the paintings "addressed me," but the speaker is "living to sit obscured by word 'here'" (*il cuore* 119). Fraser begins the poem by bracketing the deictic marker "here," a word that points to the obfuscations of this museum site and also to a particular passage in Dante (a "here" that troubled Ruskin, as we will see). In the ekphrastic exploration that follows, Fraser dramatizes tensions between error and mastery in Giotto's work through the deictic crossings of a multi-layered reading practice.

The result is a revisionary project with an underlying feminist motive: Fraser enters the Arena Chapel in the presence of male canonical writers and rearranges the textual shards of a story of male social exchanges. The story of the place is one of "paternal embarrassment" and efforts to vindicate the sins of the father though patronage of the arts, a story of fame and defamation, ritual and hypocrisy. Fraser addresses this colloquium among men: "Dante watched Giotto paint Enrico / (they talked at Arena)" (121). The colloquium extends beyond Giotto's day in the writings of Vasari and Ruskin, and she adds their texts to the evidence under consideration. The ekphrastic occasion becomes an "arena of the page"—a key figure for Fraser's feminist poetics—"a slate on which to collage and draw and *re*configure the lessons of 'the master' teacher" ("Translating" 644). As she states in an explanation of a companion poem in the same volume, "Mine is a document meant to record an alternative vision of the predominantly male archaeological point of view already well installed" ("Blank" 167). Writing this poem as if to reclaim "My Giotto" from a well-installed art-historical point of view, Fraser traverses the space for meditation, perception, and interrogation that Arena presents.

The poem opens with a figure that places the reader squarely "here," accompanying the speaker in the Arena chapel, and negotiating among the narratives of its provenance and significance:

GIOTTO

:

ARENA

:

(120)

The centered colons indicate that we are at an interface, an amplifying juncture where proper names signify an old account. We are reminded of this old account in the epigraph that immediately follows, a passage from Cary's translation of Dante's *Inferno* in which Reginaldo Scrovegni offers bitter testimony. Fraser describes him: "(money-lender of / peak avaricious habits / confirmed by cameo spot / in Dante's seventh circle)" (120). Before we see Giotto's paintings, we hear about the social circumstances that brought them about, first in Dante's voice (via Cary) and then in Fraser's re-narration. To redress his father's usurious deeds, Reginaldo's son Enrico "offers his earnest version of atonement / for paternal embarrassment" (120). He builds the chapel, dedicates it to the Virgin Annunciate as Lady of Charity, and commissions Giotto's frescoes.

After reminding us of this history, Fraser cuts to an ekphrastic glimpse, a detail from *The Last Judgment* on the entrance wall of the chapel, in which Enrico (pictured here among the blessed) hands a dollhouse-sized model of the chapel to the Virgin:

Enrico on his knees proffers
a tiny version of it
to the Annunciate, its weight
supported on another's shoulders,
salmon length of brick the same
as Virgin's gown, angel feathers'

salmon flesh and roe
lifting one swift arc (121)

Employing rhymes and slant-rhymes to describe the scene (offers / proffers / shoulders / feathers; it / Annunciate / weight; Virgin / gown /
salmon), Fraser highlights the formal symmetry of Giotto's color rhymes,
his balance of "salmon" elements. At the same time, she observes the odd-
ity of the architectural miniature depicted here: it appears weightless in
Enrico's outstretched hand, but heavy in the grip of the man to whom he
passes the burden of labor. In this opening movement with its mix of quotation and description, Fraser is simultaneously reading the tensions in the
frescoes and in the texts that describe them. The opening figure is then
echoed as she traces the history of the site:

ARENA

new name, old site, chapel built
above more than one original,
the first an amphitheatre cast
along Roman lines

ARENA   (120)

As a place where a "new name" meets an "old site," the chapel provokes a
"palimpsest dialogue" (Fraser, "Blank" 168),[12] a series of fragmentary nota-
tions that cluster into taut formal units, such as this off-rhymed *abab* quatrain, compressed with clipped "t" sounds. These notational clusters of
perceptions, meditations, and background information remind us that we
must interpret "more than one original" in this place. As Fraser's centered
words stress, we are in an "ARENA"—a field of play and of conflict.
    The word "ARENA," repeated in capitals four times in the poem's open-
ing section, is both a place in Padua and a loaded term in Fraser's poetics. In a
recent essay subtitled "Visual Poetics, as Projected through Olson's 'Field' into
Current Female Writing Practice," Fraser describes the importance of the
"arena of the page" to her feminist experimentalism ("Translating" 643).[13]
Olson, she explains, "helped to stake out an arena whose initial usefulness to
the poem began to be inventively explored by American women—in some
cases drastically *re*conceived [. . .]" (644).[14] Subsequent engagement with
French feminist theory led her to link the "arena of the page" experiments she
found in field poetics with ideas of female subjectivity:

    The dimensionality of the full page invites multiplicity, synchronicity, elas-
    ticity . . . perhaps the very female subjectivity proposed by Julia Kristeva as

linking both cyclical and monumental time. [. . .] What has been left
out of the poetic account of women in time is now manifestly present
through the developing use of the page as a four-sided document. Such
poetry focuses on the visual potential of the page for collage, extension,
pictorial gesture and fragmentation [. . .]. ("Translating" 643)

For a poet as preoccupied as Fraser is with verbal coincidence, the place
name "Arena" is irresistible for its connection to this other usage in her
poetic theorizing. Carrying theoretical baggage, it invites us into the
poem as an ars poetica, an exhibition of Fraser's characteristic method
and intent.

It is worth pausing at this point to clarify what Fraser came to see as
feminist about her formal investigations of the "arena of the page." Beginning to write and publish in the sixties and seventies, she was reluctant to
ally herself with feminist poetic positions that emerged, downstream of con-
fessionalism, as a function of "essentialist views of female language or female
poetry" (Interview 13). She found herself "chafing at the confines of the
typical 'I'-centered, mainstream American poem that so theatrically and
narcissistically positioned the writer at the hub of all pain and glory"
("Blank" 170), and sought instead a more innovative formal engagement, "a
way to move out of the lyric vise" (Interview 16). Exploring alternative
models and sources as diverse as Coleridge's marginalia, the "language tex-
tures and syntactic invention" of Stevens and Hopkins (Interview 23),
Objectivist poetry, action painting, and New York School poetry,[15] she
linked these explorations to her thinking about gender:

> Although I was increasingly conscious of gender issues, I never thought
> of participating in the construction of a separate "female" language.
> What did have meaning for me, thinking about structure, was a notion
> of "female time." While, undoubtedly, there were men who lived within
> this interruptive daily structure, it seemed to be a much more prevalent
> pattern in female life, as marked by the periodicity of professional nur-
> turing [. . .]. For many years, I've tried to find poetic gestures for
> recording broken-up time: the multiple tracks (selves) in the poem, the
> thought-lines and arguing voices we carry internally that continuously
> negotiate and interact with each other—all our fantasies of talk and
> revision and description and argument. (Interview 14)

Fraser herself names "Giotto : ARENA" as a representative result of her
"desire to trap in the poem the experience of layered, multiple perception—

not just fragmentation and interruption, but the holding-in-the-mind-of-four-things-at-once. [. . .] Layered or constellated time appeared to be characteristic of female experience but, again, not exclusively" (14,15). Presenting "thought-lines and arguing voices" that no singular point of view could contain, the poem is a record of "talk and revision and description and argument."

In keeping with my larger aim for this chapter, I offer this poem as an example of an ekphrastic poem where gender matters, but where the commonplace critical equations for gendered looking do not obtain. Tensions in this ekphrastic encounter are *not* keyed to the gender of figures or images in the works of art, nor to the gender of the beholder or author. Fraser might agree with Mitchell that ekphrasis "looks different" when written by women—but only if "looks different" is taken literally. In this poem, shaped stanzas, floating fragments, inscribed figures, horizontal lines, and typographic and freehand collages make up 21 page-fields that are also divided roughly into 11 lyric or expository movements. For example, one set of facing pages (126–7) includes these elements: 1) a symmetrical but syntactically disjunctive stanza in the manner of the Language writers, where lines such as "opponent rubied flower bend" and "intervals frame subdued" have acoustic and juxtapositional resonance; 2) a quotation from Ruskin; 3) a visual figure that inscribes a line from Dante in Italian in the center of a sketched circle; and 4) a caption that compresses and reconstitutes the previous citation. This formal, typographical, and spatial "difference" from most ekphrastic poems—this visual heterogeneity—is not the ideological difference that Mitchell and most commentators on ekphrasis have in mind. On the contrary, Fraser's efforts to theorize a feminist poetics have informed the complex *form* of this work, a work whose subject matter in no way supports an oppositional equation for tensions between male gazers and feminine objects.[16]

The tensions that Fraser dramatizes in this poem are those between "error" and "mastery" in *written* approaches to an acknowledged visual masterwork: by foregrounding the "errors" that inhere in efforts to *read* both bodies and texts, Fraser suggests ways to reilluminate Giotto's meanings. She initiates this inquiry by revisiting the idea of his "mastery," his legendary reputation for visual perfection, through two of Vasari's anecdotes: Cimabue's discovery of Giotto the shepherd boy drawing with instinctual skill on a rock, and Giotto the acknowledged master flouting the expectations of papal authority with his confidence in his technical accomplishment (Vasari 57–8, 64–5). Fraser recounts these stories in seven five-line stanzas that trace an arc away from the left margin, beginning with this elegant prologue:

rubied flower far-away bends
   at intervals
      through framework of each leaf
         sublime form's
         restrained palliate (124)

Playing on "b" and "l" sounds in an intricate stanza form, Fraser evokes a mood of homage to Giotto's "sublime form." The page is printed with the text doubled, the wing-shaped passage duplicated underneath the main text in a miniscule font as a kind of shadow/echo of this contained lyric interlude (124). Perfection and symmetry are thus both thematic and formal concerns as she addresses the story of "Giotto's O": when the "Papal courier en route scouting Vatican art among masters asks Giotto for proofs" (125), Giotto responds by offering only a perfect freehand circle, painted with a brush on the spot. Fraser highlights the double sense of "proofs": though the scout expects preliminary drafts, embellishments, and efforts to impress, Giotto offers only a single piece of evidence—one continuous mark from his "'*Pennello tinto di rosso*' / (brush dipped / in red)" (125). Ruskin, whom Fraser has already quoted, reads this famous story as evidence of Giotto's perfection: "Such a feat as this is completely possible to a well-disciplined painter's hand, but utterly impossible to any other; and the circle so drawn was the most convincing proof Giotto could give of his decision of eye and perfectness of practice" (15). Ruskin's emphasis on perfection is clear: "the most perfect power and genius are shown by the accuracy which disdains error, and the faithfulness which fears it" (16).

As if Ruskin's book were open beside her as she attempts her own account, Fraser resists his conclusion, introducing the simultaneity of "error" to counterbalance his emphasis on accuracy and perfection. She inserts typos as she quotes him—"accuracy's / disdaining errorr" (128)—and puts in stutters—"one sasaid" (130). She gives us what she calls "fFretwork" (129), finding moments of "departure" (129) from Ruskin's text. One such departure occurs when she deliberately misquotes the passage in Ruskin that follows the lines I have just cited. Here is Fraser's mis-transcription, presented in prose: "Nothing is required for the job but firmness of hand. Nothing more is said and nothing further appears to be thought of expression or invention of devotional sentiment" (128). Fraser inverts Ruskin's sentences, adds "for the job," "more," and "further," and changes the final "or" to an "of." This last change addresses Ruskin's reliance on faithfulness. Fraser imagines the "invention *of* devotional sentiment," suggesting that devotion can be concocted, summoned opportunistically for the occasion. Whereas Ruskin asks, regarding Giotto's confident "O," "Is there occult satire in the example of his art which he sends [the pope]?" (16), Fraser suggests that the answer is an unequivocal yes. She then rearranges his sentences in short-lined couplets, underscoring the possibility of satire, of challenge to religious authority, by taking "error" a step further with a pun: "Nothing's sad / nor appears / / to be thought / of devotional / / sediment" (128). Devotional "sediment" indeed: by altering Ruskin's text in this way, opening it up to typographic entropy (note also that "said" becomes "sad"), Fraser introduces a moment of irreverent resistance into official homage.

Fraser's use of Ruskin is not unlike Sherrie Levine's photographs of masterworks, in that she reproduces his text to frame its terms and suggest critique, while nonetheless leaving it basically intact and readable.[17] Ruskin, she readily concedes, has seen the "error" in Giotto all along. In fact, Ruskin's own complaint about the artist, that his faulty drawing makes his work difficult to copy, prompts her investigation of "faulty copying" in the first place. Ruskin explains, "it is by no means easy to express *exactly* the error, and *no more than* the error, of his original" (36). Fraser seizes upon this and other moments when Ruskin is concerned with error in Giotto's works and in the commentaries surrounding them. At one point he corrects a detail in Vasari, indicating that the pope in question was not "Benedict IX" but "Boniface VIII" (14), and Fraser reproduces this remark by printing a longhand "X" in her poem over the first name, giving us a palimpsest of the error and its correction. One moment in "Giotto : ARENA" self-describes this poetic preference for letting the errors show. Fraser is less interested in perfection than in "unexpected starts of effort or flashes of knowledge in / accidental directions gradually forming" (130).

Privileging error in this way, Fraser explores the traditional paradox of Giotto's greatness: he is renowned for his perfection, his consummate skill, as well as for his revolutionary ability to *break* with perfection and "type," to move beyond "cold and formal" Byzantine conventions (Ruskin 18). To stress that this "break with type" is a function of Giotto's emphasis on and involvement of the body, Fraser writes "system" and "wrist" into this passage by hand. The words in brackets appear in published versions of this poem as inked lowercase letters:

> "Not by [system], but by
> [wrist],"
> G. said,
> substituting body parts. (122)

She focuses on the central bodily detail of Vasari's account of Giotto's "O": "he closed his arm to his side, so as to make a sort of compass of it, and then with a twist of his hand drew such a perfect circle that it was a marvel to see" (Vasari 64). To represent Giotto's action, Fraser handwrites words into her lines, making signs in her own medium "not by system but by wrist." She draws attention to the bodily mechanism of both drawing and writing, the making of marks on paper through a momentary gesture of the hand.

Immediately following this passage, on the same page (122), are two meditations on bodies in Giotto's work. The first is strikingly deictic, interposing a "you" into the museum exchange between beholder and artwork: "Odd arch / of nose, / / did you notice?" This ekphrastic interrogation, a question of *atypicality* (a nose that is "odd"), is presented as a snippet of overheard dialogue, a call to a companion to pay closer attention. A similar query is posed several pages later, when the speaker asks:

> Could we trade length of dress?
> Paint unpredicted folds where thigh
> opens outward, joints resist
> (large blank surfaces) (132)

Fraser is preoccupied with moments of ekphrastic uncertainty, with visual aporias—"large blank surfaces"—and the slippery readability of images of bodies. This glimpse of a thigh—pointedly, of indeterminate gender—suggests that in his effort to represent the human figure in "unpredicted" forms, Giotto encountered both the intransigence of his materials and the sensual immediacy of the attempt. Working "not by system, but by wrist" opens the artist to the unruly energies of the body, to error and desire. To return to the visual field on which Fraser articulates this notion in longhand, we find an expanse of white space, a "large blank surface," followed by this visual-verbal summary:

> ___massed_____
> He masses pale clothed bodies—relieved with beloved and
> random Venetian stripes; blue is sparingly ppressedd . . .

(122)

Describing Giotto's innovative use of masses of color for human figures, Fraser literally *underscores* a palpable sense of physicality by stacking "massed" and "masses" in this figure. The "error" of the doubled letters in "ppressedd" reiterates this sense of pressure and physical force: the sensory appeal of Giotto's painted bodies, Fraser suggests, emerges from his evocations of flux, mass, and gesture.[18]

Fraser celebrates Giotto's "error" for its capacity to liberate sensory richness and lush particulars, an aesthetic capacity she dwells on by conjuring the
vividness of her own medium. One ekphrastic glimpse is an opportunity for
a consonantal riff: "Cypress hedges, masses of oleander, magnolia inlaid with
flutter" (123). Another affords the visual-verbal humor of "ludicrous, cum-
brous sheep" (133), and another, poignant observation:

> —four horizontally (lambs,
> too) in doorway—noting
> nature's tendency
> circle where heat lifts
>
> Gesture of damp gnawing grief (132)

In this glimpse, visual analysis of Giotto's composition (probably the famous
Pietà, though Fraser combines various details) leads to the recognition of a "gesture of damp gnawing grief." Giotto has captured "nature's tendency / to circle"
and represented this phenomenon to powerful emotional effect. Fraser describes
this effect as a function of his bodily realism, his recognition that the medieval
"great system" lacked variety and erased difference: "Real faces needed in *the
great system of perfect color,* / and different sorts of hair, G. thought" (134). Her
vivid descriptions bring out Giotto's use of botanical and anatomical detail,
departures from a "perfect" system that produce moving aesthetic effects.

An ekphrasis of *Joachim Retires to the Sheepfold,* the second scene in
Giotto's cycle portraying the early life of the Virgin Mary, also highlights
Giotto's attention to bodies:

> Joachim,
> in spite of gold-bordered cape
> and halo backdrop returns
> empty-handed, marcelled grey hair
> (curled rows). Also shepherds' mauve socks
> rolled at ankles like us.
> White dog jumps up.
> No response from Joachim,
> eYeSe sidelong.
> > > > rounder than
> > > > > O
>
> His own palpate softens theory's sharp folds (134–5)

Fraser draws attention to the way Giotto emphasizes not Joachim's haloed participation in a divine scheme, but his all-too-familiar human despondency. She circulates "l" and "r" sounds throughout this description—marcelled, curled, rolled, ankles, like—to convey the scene's centripetal unity. Meanwhile, notice of the shepherds' rolled socks provokes a deictic moment of recognition—"like us"—a tangent that points out of the frame toward the contemporary world. Fraser draws us into the process of her perceptions. The passage's strong enjambment suggests a rapid survey of details, a survey that is suddenly curtailed by the mention of the white dog. Contained in an end-stopped line, this detail punctuates Joachim's non-response. He is fixed on his purpose (making an acceptable sacrifice to the temple, hitherto rejected because he is childless), and Fraser's typographical play suggests his resolve ("YeS"). The description then circles back to her textual investigations. As Vasari relates, "Rounder than O" invokes the idiom for a simpleton that Giotto's legendary circle engendered (65) (an idiom that Fraser deconstructs, in the first edition of this poem, with a typographical pun by inserting an elliptical zero for the putatively round letter). In passages like these, where gestures, shapes, and suggestions of story enter the field of vision, Fraser emphasizes the ways Giotto's "palpate," his bodily sense of touch, "softens theory's sharp folds."

Even as she pays close attention to Giotto's bodies and the sensory appeal of his work, Fraser's field of vision remains a field of reading, a collage of verbal fragments that reminds us how inevitably we are seeing Giotto through *master texts*, through the accumulated observations and interpretations of canonical masters. We are "here" at Arena, physically present as the opening of the poem attests, but the deictic marker also functions as a bookmark holding a particular page: a passage from Cary's translation of the *Inferno* in which Ruskin identifies two problematic words—"here" and "curd." Fraser seizes upon Ruskin's two lexical quibbles, quoting the footnote in which he objects to Cary's "transposition of the word 'here'" (8n, qtd. in Fraser 127, with "word" as "world"). In the lines in question, Reginaldo Scrovegni says that his neighbor Vitaliano will soon join him in hell: Cary renders it, "my neighbor here, / Vitaliano, on my left shall sit." If one follows the Italian syntax more closely,[19] as Mandelbaum does, the line is more forcefully damning: "my neighbor Vitaliano / shall yet sit here" (153). The "here" is not merely an orienting gesture of familiarity, but an ominous benchmark—"here" is hell. Ruskin corrects the translation that would allow the deictic marker to slip to a weaker position, and then, inexplicably, worries over the word "curd." He reads it as an indication of Cary's being "afraid of the excessive homeliness" of Dante's original "butter" in the description of one of the purses worn around the neck of a usurer as a sign of infamy: "Another I beheld, than blood more red / A goose display of whiter wing than curd" (Ruskin 8n, qtd. in Fraser 119, 127).[20] Ruskin does not interpret the meaning of this "badge of shame," but notes that it appears elsewhere, "in the windows of Bourges cathedral" (8n, qtd. in Fraser 127). His footnote is curiously unrevealing, and Fraser draws attention to its odd opacity and elision. Why, for what etymological or connotative reason, would "butter" be more homely than "curd"? What about the strangeness of this emblematic punishment? What do these symbols represent, and why would likening the color of a goose to a dairy product be disturbing to Cary? Zooming in on this seemingly trivial footnote, Fraser pinpoints a moment of ambiguity in Ruskin's discussion. She extracts "here" and "curd" from this text and turns them loose in her poem, where they catalyze wordplay, one for its destabilizing deictic slipperiness (the position "here" never holds still), the other for its domestic specificity, its guttural thing-ness.

As these two words circulate in the concluding movement of the poem, Fraser suggests that a tinge of discomfort has entered Ruskin's reading, and initiated her own revisionary reading, through a hint of gendered significance in that "homely" noun. So far the poem has been marked by a significant *absence* of gendered subject matter, by a conspicuously *ungendered* attentiveness to bodies. In its final field—its final "arena" of constellated thought-lines, revisions, perceptions—the poem invites the possibility that gender registers "here":

> The widows of whiter than butter
>
> I knew none of them
>
> nor curd's buttery purse nor
>
> shame effect.
>
> Sit away, for instance, to the neck.
>
> In 'here' cathedral's obscure badge. (137)

The "widows"—perhaps the women in *The Meeting at the Golden Gate* in the fresco cycle, the wives of the usurers, or a play on Ruskin's "windows"— appear as the unspoken proprietors of the emblems of infamy that mark the condemned men. Fraser posits their existence, then doubts it. The image of a white wing on a red purse remains an "obscure badge," an elusive visual sign. If Dante's "butter" ever did carry any gendered significance, any reference to a female sphere or female experience perhaps, then that significance has been lost—it is specific to a time and place that this museumgoer cannot retrieve. The poem concludes in a deictic mode, with the here and now of a first-person speaker, an "I" who is oriented ambiguously in relation to information that has *not* been exposed: the "widows of whiter than butter" are absent from the male accounts that inform this ekphrastic encounter.

Fraser's revisionary reading has highlighted the gaps, ambiguities, and uncertainties of these accounts. Examining Giotto's famous frescoes, she has emphasized error over mastery, actual bodies over types, indeterminacy over unity, trivial detail over large significance, threads of argument over summary positions. She records this "alternative vision" to the predominantly male art-historical point of view "already well installed" by dissecting and reassembling fragments of those old accounts in a visual-verbal "arena" of wordplay and juxtaposition. Making no essentialist claims for reading or seeing "like a woman," she asserts a difference at the level of formal composition and visual page layout that allows for the representation of interruption, multiple tracks of argument, and layered perceptions. With its disjunctive strata, "Giotto : ARENA" moves through a location that is both physical—the space of the chapel-qua-museum where Fraser stands "in the actual presence of Giotto's work"—and textual—the space of reading that provides the cultural context for that work. The poem ends with "here" in scare quotes, pointing us to the Arena chapel, to Reginaldo Scrovegni's neighbor, and even to Bourges cathedral, because Ruskin directed us there. Fraser's "here" refuses to hold still, inviting us into the "arena" of her reading only to diffuse its matrix of reference and resonance.

"A WOMAN, AS USUAL, IS THE PROBLEM": ANNE CARSON'S ANTI-MUSE

"Canicula di Anna" opens with the subtitle "What Do We Have Here?" Like Swensen and Fraser, Carson launches an ekphrastic project by amplifying deixis: "Here" points to a particular *topos*—a place and topic—in which "we" and the speaker find ourselves. She too foregrounds the background of the ekphrastic encounter, beginning the 41-page poem by documenting the geographical and historical coordinates of a site-specific engagement with art:

What we have here
is the story of a painter.

It occurs in Perugia
(ancient Perusia)
where lived the painter Pietro Vannucci
(c. 1445–1523)
who was called Perugino [. . .]. (*Plainwater* 49)

Carson's mode of telling is more essayistic than Swensen's or Fraser's; she is less suspicious of narrative, standard syntax, and authorial presence, and more dismissive of the postmodern fetishization of the fragment. At the level of the sentence and verse paragraph, she is more direct, bringing to her poetic work some professorial didacticism:[21]

What do you need to know?
There are a few things.
In the fifteenth century
the Dukes of Perugia,
besieged by the forces of the Pope,
withdrew within the rock on which their city was built
and established a second,
interior city.
It came to be called La Rocca
and it did not save them
but it is still there.
The air inside it is astonishingly cold. (49)

Carson differs from both Swensen and Fraser in maintaining this expository lucidity, but she nonetheless shares their experimental, interrogative stance— an attitude of 'let's see where this leads us as the words unravel.' In her work as in theirs, this stance yields collages of chaotic quotation, imagistic riffs, tangential storylines, miscellaneous lexicons, and meandering refrains.[22] In "Canicula di Anna," Carson pieces together 53 short sections to produce a disorienting sequence that is simultaneously historical and contemporary. Like Swensen and Fraser, she stages this simultaneity by directing us to a place that *represents* the past to the present—a museum, sepulchral but startling.

In this case it is the tourist site known as Rocca Paolina, a labyrinthine fortress built by Pope Paul III between 1540 and 1543 over the medieval houses where Perugian nobility had resisted papal armies. It signifies the final defeat of the Baglioni family, who had ruled Perugia as an independent city and who had commissioned much of Perugino's work (Banker 43). Turning to this "subterranean world" (Johnstone 207) where the conqueror's fortress preserves the art of the conquered, Carson stresses that it is "still there"—a deictic indicator positions her description in an ongoing present tense of reference. She highlights the coexistence of old and new in this place, a building where escalators enable exploration of ruins that double as exhibition spaces: "In the galleries above La Rocca, / concerts are held nowadays. A pale green dead Christ / regards the cellist" (71). Through juxtapositions like this one, where medieval imagery meets modern entertainment, Carson sets up an odd poetic scenario based on simultaneity and anachronism. The painter in question turns out to be both Perugino and a late-twentieth-century counterpart, or reincarnation, pursuing his elusive muse at a phenomenology conference in Perugia. We learn in the opening section that the philosophers "have commissioned, / for the purposes of public relations, / a painter to record them / in pigments of the fifteenth century" (50), and Perugino is the man for the job. He achieved fame in his lifetime, Carson notes, through his innovative perspective—"he applied / the novel rule / of two centers of vision" (77). She employs an analogous technique in the poem by matter-of-factly conflating Renaissance and contemporary perspectives, giving us not two distinct characters, but two temporal "centers of vision" that emerge from the same site.[23]

After introducing this setting and dual perspective, the opening section of the poem concludes by lighting the fuse of a familiar gender plot:

The painter, at any rate,
is not a happy man.
A woman, as usual, is the problem.
She too has a face
and a past
worth painting.
Does that look like enough for a story?
Vediamo. (50)

Carson sets her narrative in motion by entertaining a misogynist cliché—"A woman, as usual, is the problem"—a story of a man's unrequited longing for a mysterious, beautiful woman with "a past." But the initiating gesture is interrogatory—*we'll see*—an invitation to see what happens "here." In this section, I argue that "what we have here" in Carson's poem is an investigation of this question: Where is a woman located in an ekphrastic exchange? Where—in what visual, aesthetic, historical, cultural, and social location—is "Anna"? Carson's offbeat premise—a fifteenth-century painter looking for his muse at a late-twentieth-century academic conference in Italy—prompts a series of intermittent ekphrastic observations, all of which circle around Perugino and the museum site of La Rocca. As we trace Anna's steps through this discursive and historical site, we discover a volatile, protean figure who moves among the roles of authorial alter-ego, protagonist, object of desire, painted figure, academic philosopher, feminist heroine, and medieval victim. In the process, the poem finds its way into cul-de-sacs of contradiction over ekphrasis and gender, contradictions that Carson exercises in plotlines at once inevitable and unresolved.

These plotlines find their humorous anticlimax in a museum anecdote near the end of the poem. Before I enumerate and analyze Anna's roles, I turn to this cameo appearance by a woman who is neither muse nor feminist:

> On the last afternoon of the conference
> the phenomenologists visit La Rocca.
> They are surprised
> that there is no entrance fee
> for such a tourist attraction.
> Then, by the cold inside.
> Ruts of the rock breathe dark red air
> from the fifteenth century at them,
> making it hard to light cigarettes.
> The phenomenologists cluster
> in a passageway,
> arguing
> a point of *Dasein* from
> the morning's seminar.
> Some exit by a wrong door,
> tumbling in sudden light.
> A small phenomenologist
> from Brussels
> plans to write an article about the place
> for a publisher in New York. He is eager
> to ask the curator about the mirrors,
> and about hoarding,
> but the question
> is not understood.
> (She perceives him
> to be inquiring

after the difficulty
of piloting herself about La Rocca
on high-heeled shoes, such as she
is wearing.
She assents vigorously. "*E difficile.*") (83–4)

The prose-like speed of these lines befits the phenomenologists' practicality
and discomfort with sublimity.[24] The most sensorily heightened line of the
passage—"Ruts of the rock breathe dark red air," with its six-fold repetition
of "r" sounds and driving monosyllables—is quickly undercut by the ordinary detail of fumbling with cigarettes. One of the academics attempts to
pursue the place's history, and his interest is signaled by the richer sounds of
off-rhyming words (eager, curator, mirrors, hoarding), but most of these
museumgoers are preoccupied with their own institutionalized discourse
about being ("a point of *Dasein* from / the morning's seminar"). We might
expect that the woman whom the "small phenomenologist" questions will
challenge this apathy and initiate him into La Rocca's mysteries, but she is no
sibyl—what she utters is an uncomprehending lament of less-than-sensible
footwear. She may be "piloting herself," a reflexive verb construction that
suggests autonomy, and she may be the custodian of this cultural heritage as
curator, but she is tottering on "high-heeled shoes"—a metonym, more than
any other article of clothing perhaps, for capitulation to a sexist ideal. Carson's parenthetical aside captures a woman in a contradictory stance—
knowledgeable and authoritative, yet frivolous, a caricature of cultural
trappings.

Carson's strategy is to imagine female characters in "difficult" situations
that take gender roles to extreme endpoints, and to explore the contradictions that such attenuations expose. No other character in her oeuvre
demonstrates this strategy at work more than Anna, who appears from the
beginning as an authorial experiment: "I think that I would like to call her
Anna" (50). Carson is fully aware of the deictic indeterminacy of the first-
person pronoun—the "I"s in this sentence refer both to the painter and the
poet-speaker devising the plot.[25] Arriving at the conference registration desk,
the speaker then forgets that he/she has just invented Anna two lines before,
and becomes distraught at her absence: "I could not find / Anna's name on
the list" (50). The conference organizers are understandably perplexed, and
the speaker yields: "They do not know her here. That is, / I am free to invent
her!" (52). Anna is thus both a fictive character, and a fiction within the fiction. The speaker's euphoria about her anonymity suggests that the freedom
"to invent her" is also freedom of self-invention: "Anna," as the Italian form

of "Anne," suggests an exotic version of Carson's poetic self. Chris Jennings observes that Anna resembles the character Anna Xenia in "The Fall of Rome: A Traveller's Guide" (*Glass, Irony, and God* [1995]). In this earlier poem, "Carson's speaker projects her own self-examination onto a generalized 'stranger' [. . .]—and the word 'stranger,' as one possible translation of *xenos*, furthers the impression of symbiosis between the speaker and Anna" (Jennings 931). Similarly, in "Canicula di Anna," the first of many roles into which "Anna" is cast is that of a doubled, projected self.[26]

Identified in this way with the speaker, Anna becomes the emotional center of the poem, the locus of subjective passions and psychological urgency. The title makes clear that she is the poem's protagonist, even as its translation is not straightforward. "Canicula" is "a first-declension Latin noun meaning 'a little bitch,' in either the naive or abusive sense" (Jennings 931), and it is nearly homophonic with the Italian "canícola," meaning "dog days." Thus the title suggests "Anna's Dogs" or "Anna with Dogs," as well as "Dog Days of Anna" and "Anna the Bitch." The last of these possibilities keys the story to the painter's bitterness about a woman who is "a problem," but all suggest that the story revolves around Anna's personality and perceptions. The dog motif, which provides a nightmarish undercurrent to the decidedly undramatic setting of an academic conference, also arises from her psychological portrait. We learn that she "slept in a fever of dogs" (52), and that the quiet is penetrated by "Barking like a long scald" (65). In one hallucinatory moment, "Wild dogs, mouths dripping with such bloody / syllables, ebb and run over the ocean / floor down there" (51). As titular subject, Anna circulates among these vivid synaesthetic envisionings. It is her recurrent dream that prefigures and encapsulates the underlying tensions that drive the poem: "She is in a room, / and she is trying to close the door. / Arms and legs are forcing their way in. / Violent as lobsters" (55). The poem's urgency and strangeness arise from her point of view.

At the same time, Anna plays the opposite role: she is the absent muse, the "fram[ed] silent, beautiful, distant female object of desire" (DuPlessis 71). She occupies the position traditionally assigned to the figure of a woman in the ekphrastic confrontation, as Heffernan explains it—"the voice of male speech striving to control a female image that is both alluring and threatening" (1). The painter describes his obsession:

> I hunger for Anna.
> Passing a cubicle I saw a painting of her
> and removed it to my room,
> feverishly.

It proved to be a still life
of apricots and aqua minerale.
The glass has a crack
but it reminds me
of the times of day
at which she got hungry. (53)

"Anna" is the grammatical object of the painter's longing, and then, as object
pronoun, a figure in a painting. Desire has so colored the painter's perceptions that he mistakes a still life for a portrait: he sees the shape of her body,
but it turns out to be "apricots and aqua minerale," static items of sustenance. The painter's hungry gaze is thus literally objectifying, but he realizes
his mistake when a "crack" in the visual representation sparks a memory of
Anna's own hunger. She is both the living person and the visible thing that
provokes his desire. When he assumes his historical guise as Perugino, she is
the source of inspiration because she is an object that emits light:

The artist
mixes a color. Without
the beautiful white throat of Anna
it would have been too dark
to paint
inside La Rocca. (67)

The painter works in present tense ("mixes a color"), but the conditional
construction "would have been" slides the action into the space of the past
that La Rocca represents. In that space, an artistic effort relies on the presence of a beautiful woman's body—a "throat" that functions not as organ of
speech but as lamp. "Anna" is once again the object of a preposition, defined
as silent object.

In a strange ontological shift, however, the search for Anna leads the
painter not to an embodied presence but to a *representation*:

Perhaps Anna will come today.
A train arrives at half past three.
Once she telephoned
from somewhere in the north.
I had not seen her since winter.
(In the painting, not white for
the snow, just bluish marks.) (54)

Narrating haltingly from within a deictic present ("today"), the painter waits for Anna's arrival, but our attention to the present tense of the story's action is suddenly deflected, in the parenthesis, to "the painting"—the definite article suggests the existence of a particular work in which "bluish marks" signify snow. The narrative dimension of the poem appears to have a corresponding "painting" that can be read through ekphrasis: "Two things happened / (in the painting, a superposition of colors) / at once, both impossible. / I heard someone call Anna's name. I saw the sea" (51). Recounting uncanny occurrences (Perugia is landlocked), the painter identifies a parallel painterly effect, a "superposition of colors." As the poem proceeds, the narrative and ekphrasis continue to cross and superimpose. For example, a philosopher "is reaching across / a snowy tablecloth / rendered in daubs / of blue and black" (58). Later, when a conference-goer kills a scorpion, we learn that "The painter has used / colored earths to capture / the ugly stain on the windowsill" (66). Is the painter witnessing events that he seeks subsequently to capture in a painting, or, as the present perfect "has used" suggests, are we looking at an existing representation of events? We are never entirely sure which comes first, the story of a painter making a picture, or a picture.

Amidst these shifts, Anna becomes the object of a notional ekphrasis and dissolves into the painterly medium itself. Though no particular work is named, Carson appears to be describing a painting of the citizens of Perugia:

> You see each of them
> slide an eye to the left.
> To Anna.
>
> Powdered white lead for the long eyes
> of Perugino's
> creatures. (65)

This second-person observation, with its lilting "l"s, identifies Anna as one of Perugino's subjects, but Carson never fully describes the figure. Instead, she gives us close-up glimpses of his material, as here in first person: "To render the throat holes / (blackish red), I have acquired / sap of the tree *draco dracaena*" (58). The throat, later associated with Anna as we have already seen, is depicted in a medium with a legendary origin in dragon's blood, foreshadowing the drama to come. Elsewhere, Carson dwells on a color detail, this time in third person: "(For the coins / the painter appears to have used / verdigris, an acetate of copper / that produces a cool, rather / bluish tone / unless tempered with saffron / to bring it closer / to true green)" (61). At another point, in the imperative mood, she suggests a way to render Anna's fate: "For very deep red, / use vermilion, / a sulfide of mercury" (84). Some of Carson's richest writing comes in these glimpses of the painter's medium and its origins, glimpses that shift from grammatical person to person. She describes the strange glow on one body as achieved by "mixing / with cypress resin / a sulfide of arsenic / known as orpiment / for the flesh tones" (81). Playing with the sensory potential of these words—a hypnotic string of "s" sounds, repeated "en" sounds in "resin," "arsenic," and "orpiment"—she describes Perugino's vivid depictions of bodies. In these moments of describing one artistic medium and exploiting the variety of viewpoints and sonic potentialities of her own, Carson suggests that the figure of Anna, with her intensity and passion, arises from an attempt to represent the aesthetic power of a visual source.

Carson takes the idea that Anna is not embodied but representational to a further extreme in an anecdote that inverts the ekphrastic paradigm. The painter becomes impatient for inspiration to show up and goes looking for her in the most obvious place, a museum—etymologically, the "home of the muses." He finds her represented not visually but verbally:

> Graffiti covers the walls and tombs
> of the Museo Archeologico.
> I pass the time
> photographing occurrences
> of the name Anna, until
> forbidden by an official.
>
> Alter what? (76)

Carson sketches a moment of confrontation with institutional authority much like the one Swensen anatomizes in "Trilogy." But here the museum "official" oversees an exhibition space that has already been marked by the inscriptions of a rule-defying public. Implied in the stanza break, the official's objection that photographing the walls will alter them is ironic in light of the fact that graffiti have already compromised their preservation. The frustrated painter shrugs off the reprimand: denied access to a "real" Anna and even to a visual representation of her, he bides time with a task that underscores her absence—"the name Anna" is an empty signifier that refers to countless anonymous Annas who have been memorialized by their beloveds. Putting a camera anachronistically in Perugino's hands has the effect of transposing the terms of the ekphrastic exchange. By photographing a *name,* he produces a reverse ekphrasis, a visual representation of a verbal representation. Carson manipulates the terms of an ekphrastic project, using "Anna" as a sign that underscores its own unstable referentiality.

Meanwhile, these explorations of reference, ontological priority, and levels of representation occur in a story that parodies these issues as the preoccupations of philosophers at an academic conference. From the beginning, Carson pokes fun at the phenomenologists, starting with a pun on Perugia ("ancient Perusia") and "parousia" (50), which the dictionary tells us is a Platonic term for 'the presence in anything of the idea after which it was formed.' In *Being and Time,* Heidegger links this term to Aristotle's category of Being (*ousia*) and interprets it as having a temporal sense—Presence as Present (46).[27] This notion does seem obliquely to inform the poem's concerns with prefigurations, types, and the experiential present in which re-presentations are undertaken, but Carson does not take us too far into Heidegger (she demurs, "You will understand / more of that than I do" [50]). She is more interested in satirizing the "Tautologies [and] enigmata" (63) of the academic discourse to which the painter must bear witness: "There may be an overcoming of subjectivity. / There may be lunch. / Or an interrogation of lunch" (57).

As the painter works on his "special commission," painting "the philosophers at table and / on the way to Being" (58), he knows that for all their intellectual sophistication, the muse is missing:

> A phenomenologist from Louvain-la-Neuve
> is telling us what Heidegger thought during the winter term of 1935.
> There was an interrogation of art.
> There was a circle to be made.
> Anna was nowhere in question. (56)

In her absence, the dialogue is empty academese:

> The phenomenologists are in each other's way today.
> They cough, drop pencils.
> "Your question, which is an excellent question . . ."
> They smile.
> "Crucial."
> They point a finger.
> "Very crucial."
> Affection hovers.
> "But you, you know that text very well. . . ."

Chairs scrape.
"My interpretation is fourfold. . . ."
"Twofold. . . ."
"On the way to . . ."
"I would like to say, *ja,* just the opposite. . . ."
"You could?"
A door slams.
"You can, but it's wrong." Laughter.
"Our understanding of it must come from . . ."
"From art . . ."
"Ahistorically . . ." (60)

In this parody of academic debate, Carson captures the vacuousness and accusatory tone beneath the decorum. The substance of the philosophers' thinking is dropped behind the ellipses, and what is left is evidence of ordinary entropy, ambient noises, and human response. Satirizing a point of view that would understand "art" and matters of thought "ahistorically," Carson reasserts the here and now of utterance—the present tense of this social setting and its conventions for behavior and speech. The quarrel comes to a halt when a remark reminds the philosophers of bodily cravings, and of life around them: "A woman asks for matches" (60). Anna has arrived.

Embedded in the poem is a kind of academic fable: Carson suggests that the "muse"—a figure for passionate, motivating desire—has been regrettably absent from scholarship. But as soon as she proposes that Anna's arrival signifies the revitalization of academic discourse, she finds herself in a bind: her character falls into the role of sex object. Anna is a colleague among the philosophers, "discoursing with the phenomenologist from Wiesbaden" (64) and "Working on her lecture in the library" (66), but she is also a flirt: she summons "the keeper of the Husserl archives in Berlin" to kill a scorpion in her room (66), and she "goes dancing / with a phenomenologist / who is also a captain / in the military / (reserves)" (74). What is happening in the poem, I suspect, is that in experimenting with the figure of Anna, and with the gender plot of her mysterious vital absence, Carson cannot help but find the objectification and sexualization of women around every corner. If Anna represents the real, the sensory, and sensual, she is never far from being the beautiful, silent, mysterious object of desire. The half-expressed wish for the reinstatement of the "muse," for some provisional recuperation of an inspiring, feminine sensual presence, is at cross purposes with the idea that "Anna" is protagonist, colleague, and writing self. The fable resists its own moral, and the poem becomes surreal and confusing, dissolving into fragmentary glimpses and spurts of histrionic action when Anna is forced to play both sides.

As if to derail the Anna-as-muse plot and start another, Carson then presents Anna as a feminist heroine. The second half of the poem is concerned with this countermeasure, a novelistic characterization of Anna as a bold individual and agent of her own fate. She is a young woman who defies the nuns who oversee her education to become a philosopher, and a daughter who resists patriarchal authority:

> On the day Anna was married
> her father traveled from a foreign country
> to dissuade her.
> Ran up the aisle with his hands full of money,
> everyone turned.
> After the ceremony the two of them
> strode back down the aisle arguing,
> and left the bridegroom at the altar,
> unheard over the dogs.
> In the official portraits
> it remains unclear to me
> who is who. (61)

Here Anna is not a figure of silent beauty, but of defiance. In an inversion of the traditional dowry scenario, her father offers her money *not* to marry, but she refuses. The canine leitmotif resurfaces to herald Anna's intense psychology, and she is portrayed as acting and "arguing" on her own behalf while the bridegroom is left behind. But even in this storytelling mode, Carson's characters slip in and out of levels of representation. The end of this passage suggests an ekphrasis, as if the speaker were perusing a wedding photograph that captured the altercation. The speaker expresses a sense of confusion that characterizes the entire poem: despite the narrative expectations that are set up, we cannot be sure "who is who." We know only where the characters are located—in and around the medieval center of Perugia, a place, Carson notes ominously, "*Senza uscita*" (82). The mode of novelistic narration takes us back to this 'dead end.'

Returning to La Rocca and the historical "center of vision," Carson casts Anna in one final role. She revisits the moment of crisis that underlies the museum site, the siege that drove the Perugians into their interior city, and imagines Anna as sacrificial victim: "the last occurrence of such a ritual / in the city and archepiscopal see of / Perugia" (86). This "ritual" appears to be a fiction of Carson's own devising. The early sixteenth century in Perugia was a period of violent factionalism, misrule, slaughter, and plague, but its histories do not record female sacrifices.[28] Carson histrionically brings past and present together by dramatizing the Perugian elders' deliberations in the language of the quarreling phenomenologists:

> Inside La Rocca
> an ancient dialectic
> is under way.
> "She is irrelevant."
> "Essential."
> "She is innocent."
> "Innocence is a requirement."
> "You exaggerate."
> "You repeat yourself."
> "She is not the sin."
> "She appears in a painting of the sin."
> "Deep in the background."
> "Dead center." (80)

Stacking this abrupt dialogue in end-stopped lines has the effect of a play script, a deliberate and farcical staging of "an ancient dialectic" about a woman's role and the logic of sacrifice. Anna appears at the center of a series of contradictions: she is both irrelevant and necessary, innocent and tainted, and then—"in a painting"—background and center. Entertaining the idea that a historical melodrama at the site of La Rocca "is under way" in the present, Carson takes us back to the deictic ekphrastic point of view from which it was imagined—the place and its urgency are "still there." Suspended in this location, Anna is once again both representative and representation, the stereotype of the innocent female victim and a figure in an unnamed work of art.

On the same page, Carson juxtaposes this melodramatic account with the artistic effort that has preoccupied the entire poem, drawing a parallel between the extreme of ritual sacrifice and the ritual of a woman's aestheticization:

> The painter chooses
> where to stand
> and the ritual
> totters forward. (80)

The poem has been launched from this premise: a painter sets up his easel, directs his fervent gaze toward a woman, and waits for the ritual of inspiration and creation to unfold, however fumbling and ineffectual it may be. In the process of exploring the vicissitudes of this ritual, Carson has raised several feminist questions: What is implied by "choos[ing] to stand" opposite an "Anna," and enacting the drama of her allure? What happens when a painter creates a muse, or when any artist or author posits a female other as receptacle of projected desires or sins? What happens when the female "other" is jettisoned or sacrificed, abandoned as "inessential"? Without proposing definitive answers, Carson invites these questions in a vividly imagined context that dramatizes the blatantly misogynist implications of the idea of the muse as "a little bitch," as well as the momentum that this idea still has. With its fits and starts, its wide mood swings from the mundane to the melodramatic, and its quirky juxtapositions, Carson's poem "totters forward" like the curator on her high-heels, decidedly out of step with most feminist poetics of our time in its flirtation with the very mode of objectification it exposes.

Yet "Canicula di Anna" has destabilized that mode. Over the course of the poem, Anna has played no fewer than eight distinct roles, each of which Carson explores by imagining the possible stories—implausible as well as predictable—that it generates. Anna functions as fictional personage and visual figment, locus of feeling and object of attention, thinking woman and sexual target, contemporary heroine and historical victim. A poem that starts out as an announced ekphrastic project, a "story of a painter" and his works, becomes the story of a muse who comes to life in circumstances at once prosaic and uncanny. This extended, elliptical ekphrasis of one of "Perugino's creatures," an attempt to describe a painter's means of evoking a woman on a canvas, evolves into a repositioning of that woman at various points around the painter, and writer, who attempt to capture her. But Anna refuses to sit for a portrait: her roles are not assimilable to one "place" for a woman in the ekphrastic exchange. Highlighting this mobility and indeterminacy, Carson's conclusion—her answer to the notion that "a woman, as usual, is the problem"—is ultimately a feminist response. In one of her few lines of direct dialogue, Anna tells the painter:

> "Do not hinge on me," Anna says.
> "If you want my advice,
> do not hinge at all." (74)

Advising him to detach his vision and artistic process from any "hinging" dependency on a woman, she abdicates the role of muse.

Throughout this study I have stressed that the museum setting presents poets with a tension-fraught site where volatile issues and points of view converge, and gender issues are among these. This chapter constitutes one intervention into the prevalent idea that gender has something important to do with ekphrasis, offering the model of deictic site specificity in place of bipolar paradigms of gender difference as a way to understand that importance. The ekphrastic poems I discuss here are site-specific because they are inextricably enmeshed in the locations—physical, institutional, historical, and textual—in which their speakers approach works of visual art. As they navigate these locations, foregrounding the backgrounds of their encounters with art, they confront gendered tensions among the images, perceivers, artists, authors, and characters they find there. For these poets, looking at art is never simply a confrontation between gazer and object, but a subjective process that is susceptible to the multiple refractions of verbal and visual media, and that is attuned to particular subjective pleasures and longings. Among the verbal resources at their disposal, deixis enables them to take a critical stance that documents the complex interrelationships of variously gendered speakers and objects. Swensen "suspend[s] the here and this" of a woman's defiant touch, isolating the gestures that comprise an effort to come into contact with a painting. Fraser investigates the gendered subtexts that have been "obscured by word 'here,'" and assembles her own constelled vision of a masterpiece. Carson imagines "what we have here" when a woman assumes various guises in a ritual of artistic creation at a historical site. Situating their poems in the exigencies of "here," amidst the histories and homages that intersect in museum space, these poets reframe ekphrasis in the flux of contemporary moments.

Epilogue

The boom in ekphrastic poetry shows no signs of abating. The mode is ubiquitous in the literary magazines, where poems inspired by the visual arts continue to reflect and assess their museum contexts. In the Spring 2005 *Ploughshares*, Stephen Gibson turns from the thrashing bodies depicted on an "enormous canvas" of the Battle of Lepanto to nearby tourists wrangling with museum regulations (37–8). In the same issue, Rafael Campo describes "Making Sense of the Currency on Line for Le Musée Picasso" (21). Sherod Santos gives us "Girl Falling Asleep in the Museum Gardens" in the April 2005 *Yale Review*, and Tom Sleigh, in the most recent *TriQuarterly*, stares with "alien wonder" at a painting by Gerhard Richter and at other museum-goers in the gallery (149–50). Ekphrasis has become not simply a subgenre but a pervasive poetic strategy. Several notable first books of poems to appear in recent years feature an ekphrasis as catalyst, centerpiece, or guiding trope. Timothy Donnelly's "Anything to Fill In the Long Silences," after a mixed media work by Julião Sarmento, mythologizes a scenario of protracted ekphrastic attention. Matthea Harvey's "Self-Portraits" map the varied topoi of lyric subjectivity in a series after Max Beckmann. The eponymous final poem in Brenda Shaughnessy's *Interior with Sudden Joy* (1999), after Dorothea Tanning, is an erotic-acoustic escapade through the painting's surreal playroom and the book's psychic territory. Several other recent collections, exclusively or predominantly ekphrastic, adapt the mode in the service of very different poetic agendas—Mary Jo Bang's *The Eye Like a Strange Balloon* (2004), Debora Greger's *Western Art* (2004), and Terri Witek's *Fools and Crows* (2003). Claudia Rankine's *Don't Let Me Be Lonely* (2004) builds an "American Lyric" around ekphrases of media images, including billboards, prescription drug labels, and stills from news coverage of racist violence.

These recent examples, like the poems I have addressed throughout this study, attest to the ways ekphrasis is inherently revisionary—a mode of seeing again. The acts of attention that ekphrasis entails lead these poets in their various ways to question the interplay of surface and circumstance, to reinvent the speaking and seeing self, to reply to their sources, to assay the terms of figuration and signification, and to weave and unweave the threads of perception and cognition that comprise aesthetic experiences. One striking convergence of these aims is Susan Wheeler's recent long poem "The Debtor in the Convex Mirror," which quotes and quarrels with Ashbery's "Self-Portrait" in a layered exploration of Quentin Metsys's *The Banker and His Wife* (1514). The poem begins with a richly detailed account of the painting's objects, an account that unfolds through an inquiry into the economic circumstances that inform this scene of a sixteenth-century banker counting his coins: "He counts it out" (65), and so does she. She then juxtaposes a latter-day scene of shoplifting, her imagery circling back to the small convex mirror Metsys has placed in the foreground: "She watched the cashier in the convex mirror" (69). Meanwhile, she records another kind of debt, incorporating phrases from Ashbery's famous poem and offering her own "logarithm" of cities—New York and Antwerp. Richard Howard, who selected this poem as the winner of the 2003 *Boston Review* poetry contest, observes that the poem concludes with "a parting glance in the car that takes us away from the museum" ("Wheeler" 54): the poem's preoccupations with vanity and commerce, with common coinages of financial and literary kinds, culminate when the speaker notices her companion's image "welling in the car mirror's arc— / my own in the hubcap hull—" (78).

Poems that cast this parting glance, that keep the museum in view as they take leave of works of art, have been the focus of analysis in this study. Examining poets' attentiveness to the ways aesthetic perception is inseparable from its institutional apparatuses has allowed for a recognition of the critical insights that contemporary ekphrasis affords. The range of these insights, from a cross-section of contemporary poetry, suggests that the boom in ekphrasis represents not an appeal to a consecrated tradition—the high arts huddling together in the museum for warmth—but an increasing need to mediate the terms of aesthetic experience and the terms of institutional critique. The museum settings that frame these ekphrastic exchanges bring this need to the fore, and prompt poets' interventions into issues central to both museum critique and postwar American poetry: tensions between artistic innovation and cultural enshrinement, vanguardism and traditionalism; problems of insularity and elitism, aesthetic autonomy and elevation; matters of contextualization, appropriation, and citation. Sidelong glances in museums' halls and galleries, backward glances at encounters with museum protocols and conventions, darting glances at objects and observers, foster a mode of heightened perceptiveness and critical insight.

Reframing ekphrasis through the lens of museum-consciousness has enabled a reconsideration of several of the commonplaces of American literary criticism in the late twentieth century, a juncture in which several oppositional logics—modernism versus postmodernism, tradition versus avant-garde, lyric versus language—reached an impasse. Paul Hoover's "The Postmodern Era: A Final Exam," which appeared in *Chicago Review* in 1999, presents a list of "true or false" statements that one suspects could be quotations from actual critical essays, exposing through them some of the paradoxes of critical discourse in the past several decades. Examples from this study could be employed to support and refute each of these items:

> There is no difference between a censorate and an aesthetic.
> [. . .]
> There is no tyranny like that of "the new."
> The best poets of the avant-garde are those who most betray its mission.
> [. . .]
> The deepest point of postmodern attention is the pause button on a VCR.
> [. . .]
> Pomposity is necessary to any aesthetic.
> [. . .]
> Disjuncture heals all wounds. (109–111)

Museum-sponsored anthologies expose the ways an aesthetic of "liminality" is, if not a censorate, then a guiding principle of literary inclusion and exclusion that poets in our era have alternately embraced and chafed against. Ashbery might be named the exemplary avant-garde heretic, simultaneously resisting and yielding to the postmodern tyranny of the new. Koch exposes the missionary zeal of avant-garde ambition. Howard speaks to the problem of aesthetic pomposity by ironizing his own participation in high-cultural ritual. Swensen and Fraser utilize disjunctive methods not as palliatives but as evidence of restless cognitive and perceptual pursuits. Carson reopens a gendered conflict by writing across a disjuncture in time and conjunction of place. All of the poets I discuss in this study pause particular frames of visual experience to study them, but they sustain their acts of attention longer than most VCRs allow, and remain receptive to the cultural environments that surround those frames. In these museum poems, contemporary poets hone in upon and worry over the critical preoccupations of their age, and in the best of them, they find avenues for critical inquiry without allowing skepticism or suspicion to cancel the vital energies they find in art. If the "postmodern era" is indeed over, these poets usher in an era in which ambivalence and irresolution are mediated in diverse ways through aesthetic vigor, through absorption, perspicacity, pleasure.

Notes

NOTES TO THE INTRODUCTION

1. For a derivation and history of the term ekphrasis, see Murray Krieger, *Ekphrasis: The Illusion of the Natural Sign* (1992). For a taxonomy of ekphrastic poetry, see John Hollander, *The Gazer's Spirit: Poems Speaking to Silent Works of Art* (1995).
2. I employ the adjective form "ekphrastic" as standard usage, following Heffernan and others, though this form is not listed in the OED. I have chosen to spell the noun and adjective, as have most critics, with the "k" of the directly transliterated form "ekphrasis," and I do not italicize the noun as a foreign word—its usage in English is prevalent.
3. Hollander distinguishes between "notional" ekphrases, descriptions of imaginary or fictive works of art, and "actual" ekphrases, the countless poems from antiquity to the present day that address particular, extant art objects (*Gazer's* 4). The latter category is of primary concern in this study, though some notional ekphrases will be addressed.
4. The seminal account of the pictorialist tradition in English verse is Jean Hagstrum's *The Sister Arts: The Tradition of Literary Pictorialism and English Poetry from Dryden to Gray* (1958). A recent study in this vein is Anthony Hecht's *On the Laws of the Poetic Art* (1995).
5. Paul's *Poetry in the Museums of Modernism* (2002) describes the ways "museums offered modernist poets methods and tropes for their poetry" (195). She considers Yeats's relation to the Municipal Gallery of Modern Art (Dublin), Ezra Pound's involvement with the British Museum and Library (London), Marianne Moore's encounters with the American Museum of Natural History (New York), and Gertrude Stein's creative process as shaped by her collection at 27 rue de Fleurus (Paris).
6. For a discussion of the "white cube" as the exhibition design that exemplifies the ideology of the modernist museum, see Brian O'Doherty, *Inside the White Cube* (1999, originally published 1976).

7. The discipline of museum studies, whose theorists and historians I refer to throughout, has informed this study in four ways. First, I have drawn on theories of museums as sites of production of cultural and national values, especially Carol Duncan's analysis of the liminality and performativity of the art museum as ritual site in *Civilizing Rituals* (1995). Less directly relevant but informative in this regard have been Foucauldian analyses of the public museum as an exhibitionary complex of spectacle, surveillance, and discipline (Tony Bennett, Eileen Hooper-Greenhill). Second, histories of the modern art museum and modernist exhibition design, as well as studies of the museum in the age of mechanical reproduction, contextualize the twentieth-century creation and promotion of modernist aesthetic ideology and the idea of the avant-garde (Andre Malraux, Mary Anne Staniszewski, Bruce Altshuler, O'Doherty). Third, postmodern analyses of the modernist museum, such as Douglas Crimp's *On the Museum's Ruins* (1995), have offered terms for critique and an emphasis on the "institutional framing conditions" (Crimp 5) in relation to which an artwork's meaning is formed (James Clifford, Mieke Bal). Fourth, the institutional histories of particular museums have provided information for readings of individual poems, as have the art-historical narratives offered by particular exhibitions and collections.

8. Robert Lowell applied the distinction between the "cooked" and the "raw" to poetry (a distinction Virginia Woolf used to distance herself from Joyce's *Ulysses*) "in his acceptance speech for the National Book Award" in 1960 (Rasula 233–4). (Lévi-Strauss would use the distinction later, in *Le Cru et le Cuit* [1964] and *Mythologiques*.) The distinction between poetry of accommodation and poetry of opposition is Robert Von Hallberg's, from *American Poetry and Culture 1945–1980* (1985). Outlaws and classics are the subject of Alan Golding's study of canons in American poetry (1995). The pun on lines from Tennyson's "The Charge of the Light Brigade," in an essay titled "Can(n)on to the Right of Us, Can(n)on to the Left of Us: A Plea for Difference," is Marjorie Perloff's.

9. Rasula's wax museum offers this portrait of the mainstream literary establishment: it is "operated by the MLA and subsidized by the nationwide consortium of Associated Writing Programs. Special galleries would be dedicated to corporate benefactors, including *The New York Times Book Review*, *The New Yorker*, *Poetry*, and *American Poetry Review*. Strolling through the exhibits the familiar figures would appear with uncanny verisimilitude—Dickinson, Eliot, Stevens, Frost, Plath—posed in that peculiarly arrested stance only waxen figures have. [. . .] The canon-building anthologies operate in a dignified precinct which is tacitly equivalent to the sanctuary space of the museum [. . .]" (1, 4). Institutional changes have since brought the opposing canon of poets (a canon associated with mid-century avant-gardes including the Objectivists, the Black Mountain poets,

and the New York School, and later with Language poetry) into the mainstream. Two poets associated with that alternative canon, Lorenzo Thomas and Steve McCaffery, were members of the executive committee of the Poetry Division of the MLA in terms ending 2004 and 2005, respectively. Charles Bernstein was a member of that committee from 1998 to 2002.

10. This poem appears in Rich's *Collected Early Poems: 1950–1970* (Norton, 1993), and is used from that text by permission of the author and W. W. Norton & Company, Inc.

11. This poem appears in *John Wieners: Selected Poems 1958–1984*, edited by Raymond Foye (Santa Barbara, CA: Black Sparrow Press, 1986), and is used from that text by permission of Black Sparrow Books, an imprint of David R. Godine, Publisher, Boston.

12. The concerns here about gender and public poses foreshadow themes that became increasingly important to Rich and Wieners, both of whom were later active in gay rights movements.

13. Kathleen Fitzpatrick coins the term "anxiety of obsolescence" in her discussion of novels in the age of computers (523). The use of the phrase "huddle together for warmth" to describe the relation of the "high arts" of poetry and painting was suggested to me in conversation with W. J. T. Mitchell at the Institute of Fine Arts, New York University, in September 2000.

14. For a discussion of Henry Clay Frick's establishment of the museum as a way to be remembered as other than "the epitome of the heartless capitalist" who infamously broke the 1892 strike at the Carnegie plant in Homestead, Pennsylvania, see Duncan (72–77).

NOTES TO CHAPTER ONE

1. To facilitate examination of the museum anthologies that are the subject of this chapter, each parenthetical citation indicates the title of the anthology and the poem's pagination in it. Additional information about the poem's prior or subsequent publication in other collections, and about copyright ownership, is given in notes, as well as on the copyright acknowledgments page. The anthologies' titles will be abbreviated as follows: *With a Poet's Eye*, ed. Pat Adams (*Poet's Eye*); *Transforming Vision: Writers on Art*, ed. Edward Hirsch (*Transforming*); *A Visit to the Gallery*, ed. Richard Tillinghast (*Visit*); *Words for Images: A Gallery of Poems*, eds. John Hollander and Joanna Weber (*Words*).

2. Hayden's poem appears in *Collected Poems of Robert Hayden* (Liveright), and is used from this text by permission of Liveright Publishing Corporation.

3. Duncan's concept of "liminality," which I employ throughout this chapter in conjunction with Bourdieu's concept of "charismatic ideology," is a critical figure for a cluster of assumptions about aesthetic experience that is germane to the modern art museum setting—the experience of art is revelatory,

autotelic, intuitive, culturally secluded, and politically disengaged. In defining the term as applicable to the ritual frame of the museum, Duncan draws on the work of folklorist Arnold van Gennep, anthropologist Victor Turner, and Louvre curator Germain Bazin. In this passage, she is quoting the Swedish writer Goran Schildt.

4. The Tate anthology, which includes primarily the work of British poets, is the one exception to this study's focus on American poetry. I include it because it is of a piece with the other museum anthologies published in this period, and because of the overlap in these poets' uses and developments of the ekphrastic tradition in English verse.

5. In 1986, in addition to *With a Poet's Eye*, the Tate produced *Voices in the Gallery*, edited by Joan Abse and Dannie Abse, and *Word & Image* issued a special number of ekphrastic poems. In 1984, Milkweed Editions (Minneapolis) published *The Poet Dreaming in the Artist's House: Contemporary Poems about the Visual Arts*, edited by Emilie Buchwald and Ruth Roston. In 2001, *Voices in the Gallery: Writers on Art: Original Essays and Poetry on Works of Art from the Memorial Art Gallery, Rochester, New York*, edited by Grant Holcomb, was published by the University of Rochester Press. For another discussion of the boom in late twentieth-century ekphrasis and anthologies of ekphrastic poems, see Heffernan (135–9).

6. Sara Lundquist employs a related set of oppositional terms, "reverence" and "resistance," to discuss Barbara Guest's ekphrastic poetry.

7. The seminal account of twentieth-century poetry anthologies is Jed Rasula's *The American Poetry Wax Museum* (1996). See also Alan Golding's *From Outlaw to Classic: Canons in American Poetry* (1995), especially chapter one, "A History of American Poetry Anthologies," and Christopher Beach, "Canons, Anthologies, and the Poetic Avant-Garde" in *Poetic Culture* (1999), 82–98.

8. Beach states the typical argument succinctly: "Anthologies of American poetry published in recent years attest to a persistent split in American poetic practice. The great divide in American poetry, between a mainstream practice rooted in the professional ranks of academic creative-writing departments and an experimental or avant-garde practice of variously situated 'outsider' status, has been apparent ever since the early 1960s, but it has become even more visible in the past decade" (82).

9. These "mixed" attitudes support John Guillory's assertion, in the conclusion to *Cultural Capital* (1993), that "the experience of *any* cultural work is an experience of an always composite pleasure, Proust's 'mingled joys'" (336). Guillory raises a concern that many of these poets seem to share: if we understand that elitism underpins aesthetic experience, that "it is in the interest of the dominant classes to cultivate an 'aesthetic disposition' which endorses the conditions of restricted production" (330), how do we find or

justify another way to experience the pleasures of art? Where Bourdieu seems to reject this possibility, Guillory argues thus: "Bourdieu is certainly right that it is impossible to experience any cultural product apart from its status as cultural capital (high or low); and even more, that it is impossible to experience cultural capital as disarticulated from the system of class formation or commodity production. There is no realm of pure aesthetic experience, or object which elicits nothing but that experience. But I shall nevertheless argue that the *specificity* of aesthetic experience is not contingent upon its 'purity.' Is this 'mixed' condition not, after all, the condition of every social practice and experience?" (336). Many poems in these anthologies testify to the existence of an "impure" or "mixed" specificity of aesthetic experience.

10. For a useful summary of Bourdieu's sociology of art and critique of conventional aesthetics, see Dunn.

11. This term for the prefaces, introductions, blurbs, epilogues, and notes that surround the "main text" of a book is Gérard Genette's: "Le paratexte est donc pour nous ce par quoi un texte se fait livre et se propose comme tel à ses lecteurs, et plus généralement au public. Plus que d'une limite ou d'une frontière étanche, il s'agit ici d'un *seuil*, ou—mot de Borges à propos d'une préface—d'un 'vestibule' qui offre à tout un chacun la possibilité d'entrer, ou de rebrousser chemin. 'Zone indécise' entre le dedans et le dehors, elle-même sans limite rigoureuse, ni vers l'intérieur (le texte) ni vers l'extérieur (le discours du monde sur le texte) [. . .]" (7–8). Loizeaux uses the term "paratext" in "Ekphrasis and Textual Consciousness," where she cites its popularization by Jerome McGann in *The Textual Condition* (Princeton: Princeton UP, 1991).

12. *A Visit to the Gallery*, a 7 x 10 3/4 inch paperback with 29 color illustrations and cut-away cover, is priced at $39.95. *Words for Images* is a 5 3/4 x 10 1/4 inch hardcover with 27 illustrations (22 color) priced at $35. *Transforming Vision* is a 9 1/2 inch square hardcover with 61 illustrations (59 color) priced at $27.95. The Tate anthology, a 6 1/2 x 7 1/2 inch paperback, is out of print. All are printed on heavy bond glossy paper.

13. For a discussion of the methodological usefulness of reading ekphrasis "in its textual social matrix," attentive to "its bibliographic and linguistic codes," see Loizeaux (78, 80).

14. In "Large Bad Picture," Bishop observes that one painter's vision of an expansive sea (her great uncle's) founders on its attempt at sublimity. Describing the ships depicted in the painting, the poem concludes with the difficulty of discerning whether the ships' motive is "commerce or contemplation" (*Poems* 12).

15. Kirchwey's poem appeared subsequently in *At the Palace of Jove* (Penguin Putnam, 2002), and is used by permission of Marion Wood Books an imprint of G. P. Putnam, a division of Penguin Group (USA) Inc.

16. For a discussion of the ways Rodin's *The Kiss* is common currency—public property—in dominant gender ideologies, see Harper, who argues that "the degree to which this particular conception of heterosexual communion has been thoroughly stereotyped" (2) is a function of its relation to conceptions of privacy (1–12).

17. See, for example, *Cowboy with Cigarettes* (1990), which ironizes the Philip Morris Company's sponsorship of a Picasso exhibition (McShine 155).

18. For speculation about how "resistance is conceptualized nowadays within the metaphysics of power and has no currency outside that fashionable and gratuitous paranoia," see James R. Kincaid, "Resist Me, You Sweet Resistible You," *PMLA* 118.5 (2003): 1326.

19. As Sandra Lea Meek observes, the university has become the primary site "for creative production, performance, and consumption of poetry" in the United States. Several institutional factors have contributed to this trend, including the rise of creative writing programs, the role of university presses in publishing poetry, the numbers of poets finding employment as professors, and the fact that (as Ron Silliman notes) college syllabi provide the largest market for poetry books in the U.S. (83–4).

20. Bourdieu describes this corollary of the "charismatic ideology": "[t]he myth of an innate taste which owes nothing to the constraints of apprenticeship or to chance influences since it has been bestowed in its entirety since birth, is just one of the expressions of the recurrent illusion of a cultivated nature predating any education [. . .]" (*The Love of Art* 109).

21. These lines from Clampitt's poem, with minor changes, appear in *The Collected Poems of Amy Clampitt* (Knopf, 1997), and are used from this text by permission of Alfred A. Knopf, a division of Random House, Inc.

22. The visual arts are an important source of inspiration for Fulton throughout her work, providing analogies in "The Priming is a Negligee" [sic] in *Sensual Math* (1995) and "Scumbling" in *Palladium* (1986), among other poems. "Close" was written for *A Visit to the Gallery* and then included, after a prefatory poem, as the opening poem in her fifth collection, *Felt* (2001). For discussions of Fulton's uses of the visual arts, see Lynn Keller, "The '*Then Some* Inbetween': Alice Fulton's Feminist Experimentalism," *American Literature* 71.2 (1999): 311–340; and Cristanne Miller, "'The Erogenous Cusp,' or Intersections of Science and Gender in Alice Fulton's Poetry," in *Feminist Measures*, eds. Lynn Keller and Cristanne Miller (Ann Arbor: U of Michigan P, 1994): 326 ff.

23. Fulton's poem, with minor changes in punctuation, indentation, and stanza breaks, appears subsequently in her collection *Felt* (Norton, 2001), and is used by permission of W. W. Norton & Co, Inc. The quotations that appear here reflect the anthology version.

24. Fulton mulls over several of the issues of aura, authenticity, and exhibition raised in Walter Benjamin's "The Work of Art in the Age of Mechanical

Reproduction." She especially engages the terms of his observation, in an endnote, that "[t]he definition of aura as a 'unique phenomenon of a distance however close it may be' represents nothing but the formulation of the cult value of the work of art in categories of space and time perception. Distance is the opposite of closeness. The essentially distant object is the unapproachable one. [. . .] The closeness which one may gain from its subject matter does not impair the distance which it retains in its appearance" (243n5). In its meditations on distance and closeness, "Close" suggests that a *collapse* of distance, an approach to art without ritual distance, allows for fuller appreciation of its aesthetic value.

25. Critical discussion has focused on Wright's visionary poetics, even as critics note that "Wright's poetry is not theological or even confidently transcendental. The Christian negative way merges with modernist skepticism and negation" (Costello 325, 329). Wright's ekphrastic poems about Cézanne, Mondrian, Rothko, and Morandi are usually read as a means to creating a "symbolist poetry [that] has the edge of allegory without a key" (Costello 329), but this emphasis may be more a function of the critics' interests than Wright's own. See Helen Vendler, "Charles Wright," in *The Music of What Happens* (Cambridge, MA: Harvard UP, 1988): 388–97; Bonnie Costello, "Charles Wright's *Via Negativa:* Language, Landscape, and the Idea of God," *Contemporary Literature* 42.2 (2001): 325–46; and Edward Hirsch, "From 'The Visionary Poetics of Philip Levine and Charles Wright,'" in *The Columbia History of American Poetry,* ed. Jay Parini (New York: Columbia UP, 1994): 777–805.

26. Wright's poem appears subsequently in *Negative Blue: Selected Later Poems* (Farrar, Straus and Giroux, 2000) and is used from this text by permission of Farrar, Straus and Giroux.

27. Heffernan notes: "In the Western world, the oldest recorded comment about the difference between painting and poetry is the one Plutarch attributes to Simonides of Ceos (ca. 556–467 B.C.)—that 'painting is mute poetry and poetry a speaking picture'" (49).

28. Kunitz writes about an artist he knew personally: "Of all the artists I have been close to through the years, Philip Guston was unquestionably the most daemonic" (qtd. in Balken 65). Kunitz collaborated with Guston on several "poem-pictures," including a joint commission from the Smithsonian Institution "to design a poster celebrating the marriage of poetry and painting" in 1976, when Kunitz was Consultant in Poetry to the Library of Congress (65). Several of these collaborations are included, along with "poem-pictures" Guston made with Clark Coolidge, Bill Berkson, and William Corbett, in *Philip Guston's Poem-Pictures,* ed. Debra Bricker Balken (Seattle: U of Washington P, 1994).

29. Kunitz's poem appears subsequently in *Passing Through: The Later Poems New and Selected* (Norton, 1995), and is used from this text by permission of W. W. Norton & Company, Inc.

NOTES TO CHAPTER TWO

1. References to Ashbery's books in this chapter will be abbreviated as follows: *The Mooring of Starting Out: The First Five Books of Poetry* (MSO), *Self-Portrait in a Convex Mirror* (SP), *Houseboat Days* (HD), *As We Know* (AWK), *A Wave* (W), *April Galleons* (AG), *Flow Chart* (FC), *Hotel Lautréamont* (HL), *And the Stars Were Shining* (SWS), *Can You Hear, Bird* (CY), *Wakefulness* (Wk), *Girls on the Run* (GR), *Your Name Here* (YN), and *Reported Sightings* (RS). For information about permission to use this quoted material, see the acknowledgments and copyright pages.

2. For discussions of Ashbery's uses of the visual arts, spanning the decades, see especially Leslie Wolf, "The Brushstroke's Integrity: The Poetry of John Ashbery and the Art of Painting" (1980); Charles Altieri, "John Ashbery and the Challenge of Postmodernism in the Visual Arts" (1988); and David Sweet, "'And *Ut Pictura Poesis* Is Her Name': John Ashbery, the Plastic Arts, and the Avant-Garde" (1998).

3. David Kellogg, usefully mapping contemporary American poetry in terms of four poles of allegiance rather than the usual two (self, tradition, innovation, and community, where most critics group the first two and the second two terms together), observes that "Ashbery is claimed by powerful critics for all four poles: Helen Vendler views him as a poet of the self, Harold Bloom as a reviser of tradition, Marjorie Perloff as a formal innovator, and John Shoptaw as a writer of 'encrypted' gay poems" (103).

4. See, for example, Fred Moramarco: "With the publication of *Flow Chart* in 1991 and *Hotel Lautréamont* in 1992, John Ashbery's work appears to have come full circle. The books revert to the kind of disjunctive language characteristic of his earlier work" (38).

5. For other discussions of Ashbery's reception, see John Koethe (83–88), and Andrew Ross, "Taking the Tennis Court Oath" (193, 201–2).

6. I am indebted to the work of two critics who have also challenged the distinction between mainstream Ashbery and his avant-garde counterpart, even as I depart from their approaches in various ways. Nick Lolordo, whose useful discussion of Ashbery's reception I have drawn on here, traces a "dual conversation" in *Flow Chart* (1991), arguing that "Ashbery positions himself both in the present tense of writing (the moment of writing in 'open form' that puts him in recognizable continuity with radical contemporary practice) and within literary history (by means of an allusiveness that puts his poem in problematic contact with ideas of high modernism, specifically those of Eliot and the New Criticism)" (755–6). Yet by pointing to this "dual conversation," Lolordo upholds the distinction between radical practice and literary tradition, postmodernism and modernism, that the "two-Ashbery" logic relies upon (the word "problematic" signaling his inclination toward the former). My aim is to show how

Ashbery continually complicates the oppositions on which Lolordo's argument of duality is based. Conversely, where Lolordo argues for the copresence of *both* poles in such a way that this opposition remains intact, James Longenbach argues for *neither*. Longenbach also attempts, as I do, to revise the "narrative most often employed to characterize postmodern poetry (tradition versus innovation)" ("Ashbery" 107). He contends that "[t]he very distinction between a mainstream and an avant-garde—between the academics and the Beats, the New Critical and the confessional—seemed meaningless to Ashbery in the mid 1960s" (103). "Meaningless" is troubling here. Ashbery, consciously involved in an innovative project throughout his career, no more steps out of the game than he steps in as referee.

7. Ashbery began working as an art critic in the late fifties, writing frequently for the Paris *Herald Tribune, ARTnews, Newsweek,* and *New York* over the next three decades.

8. This prose translation is James Hynd's, included in Horace, *The Art of Poetry,* verse translation by Burton Raffel (Albany: SUNY P, 1974).

9. Ashbery's most recent books reflect the continuing influence of the visual arts on his poetry. In *Chinese Whispers* (2002), he addresses visual sources as varied as cameras obscuras and pictures on a cocoa tin ("A Sweet Place"). Other poems in the collection include "View of Delft," "Portrait with a Goat," and "The Evening of Greuze." *Where Shall I Wander* (2005) includes the poem "The Red Easel," as well as a reference to Corot in "Well-Lit Places."

10. Darger's novel, *The Story of the Vivian Girls, in What is Known as the Realms of the Unreal, of the Glandeco-Angelinnian War Storm, Caused by the Child Slave Rebellion,* was found in his apartment, with "scroll-like watercolor paintings on paper" after his death in 1973 (Schjeldahl 89). See also Thévoz (15ff).

11. See Peter Bürger, in *Theory of the Avant-Garde:* "Since now the protest of the historical avant-garde against art as an institution is accepted as *art,* the gesture of protest of the neo-avant-garde becomes inauthentic" (53).

12. Other mainstream anthologies in which Ashbery appeared include Hollander's *Poems of Our Moment* (1968), Strand's *The Contemporary American Poets* (1969), and Untermeyer's *50 Modern American and British Poets* (1973) (Rasula 488–9). Ashbery was included in Donald Allen's *The New American Poetry* (1960), and Lolordo notes the significance of Alan Golding's recent discovery that Ashbery "was the last poet *cut* from Donald Hall, Robert Pack, and Louis Simpson's *The* [sic] *New Poets of England and America;* his presence therein would have made him the only poet enlisted by both sides in the original 'anthology wars'" (751n).

13. I am sympathetic to the approach of other critics who have argued for the force of the avant-garde in twentieth-century poetry without falling into the trap of arguing for an "authentic" experimental stance. Libbie Rifkin revisits

the academic co-option of postwar poetic avant-gardes ("tenured radicals") to argue for the coexistence of "institutionalization and breakthrough," rather than "institutionalization as a necessary and final fall from grace" (129–30).

14. He is writing about Jean Fautrier, an abstractionist André Malraux claimed (in 1959) to be an "antecedent" to the New York School. Ashbery comments on avant-garde one-up-manship in terms that bear interesting relation to his own revisionary view of things a few years later. Some artists, he explains, pushed back composition dates "to prove they were slinging housepaint around when Pollock was a pup. Even if some of Fautrier's Abstract Expressionist paintings precede those of Pollock & Co., it is extremely doubtful that anyone in New York was aware of them [. . .]" (RS 136).

15. The book these lines echo most forcefully is Bishop's *Questions of Travel* (1965), which postdates this poem. Even so, it is likely that Ashbery, an avid reader of Bishop's work, knew the poems from their earlier magazine publication (which often occurred for Bishop many years before book publication), and even if he did not, he unmistakably echoes Bishop's style in these lines. For full discussions of the affinities between Ashbery and Bishop, see James McCorkle, "The Demands of Reading: Mapping, Travel and Ekphrasis in the Poetry from the 1950s of John Ashbery and Elizabeth Bishop," in Diggory and Miller, and *Poetry and the Sense of Panic: Critical Essays on Elizabeth Bishop and John Ashbery,* ed. Lionel Kelly (2000).

16. For a discussion of the range of Ashbery's allusiveness in this poem, see McHale (565).

17. In the 31 years since its first publication, the poem has invited near constant commentary, with each new interpretation registering changing critical preoccupations. In 1984, Richard Stamelman offered an account of the poem's specular and speculative deconstructions. In 1986, Lee Edelman followed up from a Lacanian perspective that exposes the poem's mise-en-abîme of subjectivity. In 1993, Stephen Paul Miller brought Marxist and New Historicist methodologies to bear on the poem, reading it (ingeniously, via Jasper Johns), as a tropological engagement with the "surveillance mechanisms" of the Watergate affair.

18. Heffernan, one of the few critics who has commented on the museum setting in this poem, describes the ways "Self-Portrait" is more laden with the museum's interpretive machinery than any previous ekphrasis: "the museum begins to play a part in ekphrastic poetry at least as early as Keats's sonnet on the Elgin marbles in 1817, but its part remains silent until Auden's 'Musée des Beaux Arts' of 1938. And likewise silent—up to 'Self-Portrait'— is the role played in ekphrasis by art-historical commentary and reproductions" (171).

19. Vendler's essay, "Reading and Hearing John Ashbery's 'Self-Portrait in a Convex Mirror,'" is printed on three sides of the record's jacket cover and is unpaginated.

20. Further citations of "Self-Portrait in a Convex Mirror" in this chapter indicate page numbers from *Self-Portrait in a Convex Mirror* (Penguin, 1976).

21. Heffernan also draws a connection to Benjamin here (172).

22. Ross Leckie and Robert Mueller both end with these lines. Richard Howard, Charles Altieri, and James Longenbach also give Ashbery the reverberating last word, ending on other quotations.

NOTES TO CHAPTER THREE

1. In "Seasons on Earth" (1987), a preface to the re-publication of two early long poems, Koch offers this ottava rima reminiscence of the academic climate of the fifties:

   > *The Waste Land* gave the time's most accurate data,
   > It seemed, and Eliot was the Great Dictator
   > Of literature. One hardly dared to wink
   > Or fool around in any way in poems,
   > And Critics poured out awful jereboams [sic]
   > To *irony, ambiguity,* and *tension*—
   > And other things I do not wish to mention. (310)

   "Jereboams" is a punning hybrid of "jeremiad" and "jeroboam," a five-liter container for wine.

2. Further citations of Koch's poems in this chapter, unless otherwise noted, refer to *On the Great Atlantic Rainway.*

3. For discussions of the New York School poets in relation to literary sources, see McCorkle and Pelton, in Diggory and Miller. See also Perloff, *Frank O'Hara,* especially chapters two and three.

4. Gauloises is a brand of cigarettes, Bonnard was the French Intimist painter (1867–1947) known for his use of color in domestic interiors, and Brendan Behan (1923–1964) was an Irish revolutionary, satirist, and playwright.

5. Koch attended Harvard after serving in the army, and Howard attended Columbia. Koch went on to graduate study at Columbia, Howard at the Sorbonne. Koch was a professor in the Columbia English Department for several decades. Howard is currently a professor in the Writing Division of Columbia's School of the Arts. Koch, in recognition that many considered overdue, received the Bollingen Prize in 1995 and was elected to the American Academy of Arts and Letters in 1996. Howard received the Pulitzer Prize for *Untitled Subjects* in 1970, the American Book Award for his translation of

Baudelaire's *Les Fleurs du Mal* in 1983, and a MacArthur Fellowship in 1996.

6. Humor in ekphrastic poetry is an underdiscussed subject. Loizeaux acknowledges its importance in her analysis of Paul Durcan's *Crazy About Women* (1991), a book of ekphrastic poems produced in conjunction with the National Gallery of Ireland. Among the strategies that Durcan uses in his refusal to ally himself with "the elitism associated with the museum world of money and power," humor is one of the most effective. "Simply being funny" is a way to disrupt the museum's cultural authority (80–91).

7. Postmodernism exiles these modernists for different reasons: Kandinsky for linking nonfiguration with spirituality, Arp for seeing a direct channel from Fantasist sculpture to the subconscious, Valéry for insisting that the work of the poet be as formally disciplined as science, Léger for his associations with formalism as Purism, and Marinetti for his extension of futurism into fascism.

8. For a detailed reading of this poem and a discussion of Guest's ekphrastic strategies more generally, see Sara Lundquist, "Reverence and Resistance: Barbara Guest, Ekphrasis, and the Female Gaze," *Contemporary Literature* 38.2 (1997): 260–86. For a discussion of Guest's collaborations and correspondence with painters, see Lundquist, "Another Poet Among Painters: Barbara Guest with Grace Hartigan and Mary Abbott," in Diggory and Miller. For Guest's own resistance to the term "postmodern," see her interview with Mark Hillringhouse (26).

9. Guest's oppositional strategy is, as Lundquist argues, also a feminist one: although "red okays adorn her," the figure Guest invokes is far from acquiescent with her "abnormal / loquaciousness." Lundquist makes the important point that "[a]s the only woman poet of the first-generation of New York poets, Guest may have seen in Miró's painting a visual validation of the inventive, humorous, fluid, elusive, urbane, whimsical, postsurrealist work she and they were doing and rejoiced to find it labeled feminine" ("Reverence" 268).

10. For a discussion of the attitudes toward history and art history represented by O'Hara's poem and Rivers's painting, see Suzanne Ferguson. For a detailed reading of the poem's "inversionary rhetoric" and both it and the painting's relation to the "heroic" period of abstract expressionism, see Michael Davidson (72–4).

11. African-American painter Robert Colescott (b. 1925) added another dimension to this satire with *George Washington Carver Crossing the Delaware* (1975). Colescott reinterprets this iconic image in American history by replacing the Revolutionary War general with Carver.

12. O'Hara worked at MoMA from 1951 until 1953, and then rejoined the museum in 1955, starting as an assistant in the International Program, becoming Assistant Curator of Painting and Sculpture Exhibitions in 1960, and Associate Curator in 1965. For an outline of O'Hara's career as an art

critic and curator, see Donald Allen's chronology in *The Collected Poems of Frank O'Hara* (xiii-xvi).

13. The manuscript of O'Hara's poem is dated November 29, 1955. It was first published in *Poetry* in March 1956.

14. Koch's play was performed again, in Boston, in 1989 and 1990, in the latter production as an opera with music by Scott Wheeler.

15. For further discussion of Koch's plays and poem-plays (such as *One Thousand Avant-Garde Plays* [1988]), see Pelton (334–5), and Lehman (*Last* 207).

16. David Lehman describes the circumstances of the poem's composition (*Last* 79–80). In correspondence with me in the spring of 2002, Koch identified this poem as the one that he and Ashbery wrote at the Rodin Museum.

17. Reflecting in a 1996 interview on the ways opposition to Eliotic and Audenesque modernism was an initiating impulse behind New York School vanguardism, Koch admits, as many critics would be wise to admit, that he had "gotten used to being a little annoyed by Eliot and forgot how marvelous 'The Waste Land' was. [. . .] Eliot later was useful to 'write against.' In an Eliot-dominated poetic ambience, even the slightest sensations of happiness or pleasure seemed rare and revolutionary poetic occasions. If 'happy,' positive, excited poetry were the 'scene,' I might have been looking for the nuances of the losses and sorrows in my life for the subjects of poems" (Interview 45). Koch stresses that his contrarian stance was not chiefly an objection to Eliot's poetry itself—reading *The Waste Land* for the first time he was "spellbound" by "its vaguenesses, its solemnity, and its dissociations" (45)—but an objection to the institutionalization of its solemnity.

18. Hayden Carruth is often cited as an "heir of Eliot" who epitomized the ways literary modernism had ossified into a New Critical orthodoxy. Marjorie Perloff, for example, groups him with Karl Shapiro, Delmore Schwartz, and Richard Wilbur as exemplars of this trend (*21st-Century* 2). Louise Bogan, from an institutional standpoint, defines the literary mainstream: she reviewed poetry for *The New Yorker* for 38 years. Her poetry, with its classicism and strict formalism, also locates her firmly within dominant modernist trends. Later discussions place her in a poetic lineage that includes Edna St. Vincent Millay, Marianne Moore, Elizabeth Bishop, and Louise Glück.

19. In 1962, the year "The Artist" appeared in Koch's first book, *Thank You and Other Poems,* the Indianapolis Museum formed the Contemporary Art Society to encourage acquisition of recent works.

20. The work that Koch likely had in mind was Ernst's *Capricorn* (1948), a piece in reinforced concrete that is his largest free-standing sculpture (Spies 75). A hieratic group in which a sceptred goat sits on a square pedestal beside a mermaid, the work is considered "the largest single achievement" of Ernst's sojourn in Sedona, Arizona, where he lived after 1946 with his fourth wife, Dorothea Tanning (Russell 208).

21. Koch's imaginary artist begins to sound more and more like Christo as the poem goes on. It is possible, but unlikely, that at the time of the poem's writing Koch had heard of Christo's earliest wrapped objects, begun around 1958, but Christo did not begin his larger wrappings, like *Dockside Packages,* until 1961. Large scale landscape projects took place much later—Christo finished wrapping the Pont Neuf in 1975, the islands in the Biscayne Bay in 1983. (Fineberg 349–59)

22. See Hoover, who points to Clark Coolidge's relation to the New York School and John Ashbery's *The Tennis Court Oath* as the central text in this genealogy ("Fables" 20).

23. After reading unsolicited manuscripts for *Poetry* magazine (which receives 80,000 submissions a year) in the mid nineties, Davis McCombs described "the archetypal plot" of this kind of mainstream poetry as one where "life itself" is inherently poetic: "Here I am, looking out my kitchen window, and I am important" (Brouwer 291).

24. Chinitz concedes that in Koch's later work, "the earlier postmodern experiments are more often an object of affectionate mockery than a model for new verse. 'Days and Nights' (1982), for instance, wryly caricatures the Koch of *When the Sun Tries to Go On* [. . .]" (324). But Chinitz downplays these counterexamples in his effort to define, as his title indicates, Koch's "postmodern comedy."

25. "*The Store* was a friendly, overfilled room of hamburgers and tennis shoes, prepackaged shirt-and-tie combinations, and reliefs of Pepsi signs and sewing machines—all made of brightly enameled plaster—with a delicious vulgarity that celebrated 'the poetry of everywhere,' as [Oldenburg] put it" (Fineberg 197). See also Altshuler (212).

26. In a 1962 letter to Robert Lowell, Bishop wrote that studying a book of Bosch prints colored her perception of everyday reality and drew out the elements in her world that seemed Bosch-like. She also remarked that poetry could have a similar effect: if a poem's perspective or imagery caused the world to resemble it for a time after reading it, she felt assured of its quality (*One Art* 409).

27. Parenthetical citations in the text refer to the individual volumes in which Howard first published these poems, with their titles abbreviated as follows: *The Damages* (D), *Two-Part Inventions* (TPI), *Fellow Feelings* (FF), *Misgivings* (M), *Lining Up* (LU), *No Traveller* (NT), *Like Most Revelations* (LMR), *Trappings* (T), and *Talking Cures* (TC). Several of the poems were reprinted in *Inner Voices: Selected Poems 1963–2003* (Farrar, Straus and Giroux, 2004) and are used by permission of Farrar, Straus and Giroux. In that volume, the poems I cite begin on the following pages: "Nadar" (212), "Lining Up" (217), "Even in Paris" (247), and "Hanging the Artist" (417). Other quotations from Howard's poems that do not appear in *Inner Voices* are used by permission of the author.

28. The dedication to *Untitled Subjects* reads: "To my friend Renaud Bruce, [. . .] less presumptuously than if I were to inscribe them, as well, to the great poet of otherness whose initials are the same and who said, as I should like to say, 'I'll tell my state as though 'twere none of mine.'"

29. Howard's translations of Barthes include *A Lover's Discourse: Fragments* (1978), *The Eiffel Tower, and other mythologies* (1979), *Camera Lucida* (1981), *The Responsibility of Forms* (1985), *The Rustle of Language* (1986), *Empire of Signs* (1982), *The Semiotic Challenge* (1988), and *Incidents* (1992). He has also written prefaces for editions of *The Pleasure of the Text* and *S/Z*.

30. *Misgivings* includes the 13-poem series "Homage to Nadar," and *Lining Up* includes the nine-poem series "Homage to Nadar (II)."

31. For a discussion of Howard's historicism in relation to his construction of a cultural matrix as a gay poet, see David Bergman.

32. Ramke makes a similar comparison between Howard and Susan Howe (127).

33. Frequently cited as a watershed performance and "classic" Happening, Schneemann's "Meat Joy," a group performance at the Judson Memorial Church in New York in November 1964, was an ecstatic and repellant exploration of the flesh as material. See Carolee Schneemann, *More than Meat Joy: Complete Performance Works & Selected Writings*, ed. Bruce McPherson (New Paltz: Documentext, 1979).

34. Fineberg notes that the Happenings' anti-bourgeois stance was cancelled once they ceased to shock and started to be stylish entertainment. He observes that "by 1962 the whole phenomenon had grown too commercialized, according to Oldenburg: 'People were arriving in Cadillacs'" (191).

35. Opened in 1969, the museum is now known as the Norton Simon Museum.

36. In his essay "Fragments of a 'Rodin,'" cited earlier, Howard includes this prose description of *The Burghers of Calais*: the work exemplifies "the Rodin who scandalized the townspeople of Calais by insisting that civic virtue and patriotic sacrifice were not always noble and exalted, that heroism is a form of solitude—a suffering form" (*Paper Trail* 192).

37. For a reading of this poem's formal and thematic relations to Howard's earlier work, see Mazzaro.

38. Stevens's letters are full of references to the Paris of his imagination and the Paris he knew from corresponding with booksellers, art dealers, and friends. In 1952, he writes, after hearing about Paris from a colleague who had recently been there, "I sighed to think that it must forever remain terra incognita for me [. . .]" (*Letters* 755). In 1953, he explains, "I am one of the many people around the world who live from time to time in a Paris that has never existed [. . .] [that] may be wholly a fiction, but, if so, it is a precious fiction" (773).

NOTES TO CHAPTER FOUR

1. This famously controversial work, a 12-foot high and 120-foot long arc of leaning steel commissioned for the Federal Plaza in New York, was removed because of its disturbing massiveness: "As it was a site-specific work, this removal effectively destroyed it" (Fineberg 321).

2. Kwon traces the use of the term "site-specific" in three phases or paradigms. In the first, as exemplified in the works of Serra, Robert Smithson, and Robert Barry, a site-specific work was necessarily grounded in a tangible reality. These artists sought to challenge the modernist paradigm of the "autonomous and self-referential, thus transportable, placeless, and nomadic" work of art by emphasizing "the materiality of the natural landscape or the impure and ordinary space of the everyday," and by demanding "the physical presence of the viewer for the work's completion" (11–12). In the second paradigm, works by Michael Asher, Marcel Broodthaers, Daniel Buren, and Mierle Laderman Ukeles foregrounded the "*cultural* framework defined by institutions of art" as well as "the social matrix of the class, race, gender, and sexuality of the viewing subject": "To be 'specific' to such a site [. . .] is to decode and/or recode the institutional conventions so as to expose their hidden operations" (13–14). Third, site specificity entailed an "expanded engagement with culture [that] favors public sites outside the traditional confines of art both in physical and intellectual terms" (24): a site is thus a "discursive formation" (30).

3. For further examples of the pervasive emphasis on gendered distinctions in accounts of ekphrasis, see Katy Aisenberg and Grant Scott. Drawing on the work of Stephen Nichols, Aisenberg's *Ravishing Images* (1995) focuses on "the mastery implicit in the male gaze of ekphrasis and the consequent vulnerability of the female subject" (15). Aisenberg argues that the relationship between words and images in ekphrastic poetry is "less a marriage than a rape. Fearing the silent, wordless power of the Other, these poets actually spoke for these images in the rhetoric of possession, subordination, violence, or entombment. The attempt to fix an image in order to empower language consistently reveals the projection of the poet's own desires onto the Other and a dominating act of mastery" (1). Similarly, Scott argues that ekphrasis reflects "the anxieties of the perceiving eye, the *agon* or rivalry that emerges between the sister arts, and the persistently (though not exclusively) masculine project of surveying and attempting to control the feminized artwork" (64).

4. Another critic who has challenged these reductive gender paradigms is Sara Lundquist. She objects to the ways "most discussions of ekphrastic poetry have assumed that the viewer of the art will be male, engendering an entire literature describing the desires and the appropriations of the 'male gaze'" ("Reverence" 282). Lundquist has shown that Barbara Guest's ekphrastic

poems "speak *with* the paintings and, through them, *with* the painters; they speak relationally rather than paragonally" (284).

5. I am putting the term deixis to new use in relation to ekphrasis by applying it to these experimental poets' gender-specific engagements with museum sites, and by stressing the pervasiveness of this strategy in these poets' work, but discussion of deixis in ekphrasis is not itself new. Central to many theories of ekphrasis are discussions of the gazer's positionality, and the ways that the gazer's relative position is highlighted in the ekphrastic text. See, for example, Hollander's *The Gazer's Spirit*, especially his treatment of the importance of the "presence of a gazer, reporting what he or she sees, variously describing what is there to be seen" as a key feature of modern ekphrasis (32).

6. The relevance of Benveniste's linguistic analysis was first suggested to me by Norman Bryson's reading of Mieke Bal's revisionary art criticism in *Looking In* (2001).

7. Swensen derives her term from Gertrude Stein, whom she quotes in the essay's epigraph: "I write entirely with my eyes" (122).

8. Swensen has produced what can be considered an ekphrastic oeuvre. In addition to *Try* (1999) and *Such Rich Hour* (2001), she recently published *Goest* (Alice James Books, 2004), which is framed in sequences on Cy Twombly's sculptures.

9. Swensen shares Fraser's concern, as we will see, with the textual backdrop of her visual observations. *Such Rich Hour* draws on a wide range of source materials (the book includes a 27-item bibliography) for information about fifteenth-century culture, including discussions of hagiography, mathematics, apiculture, and other subjects.

10. Albers's study of color, originally published in 1963, presents examples and experiments that demonstrate the ways in which "the same color evokes innumerable readings" (1). He shows that "in visual perception a color is almost never seen as it really is—as it physically is," but rather appears differently depending on its context, as a result of several visual phenomena, including gradation, reversal of grounds, subtraction, after-images, and optical mixture (1, 77–81).

11. In the introduction to her interview with Fraser, Cynthia Hogue observes that "the sight of fragments" in museums "catalyzed" Fraser's poetic process in "Etruscan Pages," another poem in *when new time folds up* (1993) (Interview 3). A similar process produces "Giotto : ARENA" and other ekphrastic poems in Fraser's oeuvre, including *Magritte Series* (1977); "La La at the Cirque Fernando, Paris," after Degas; "WING," after Mel Bochner's drawings and installations; and "Cue or Starting Point," after the works of Sanda Iliescu.

12. Fraser describes H.D.'s concept of the palimpsest as an important model for her work. See her essays "The Blank Page: H.D.'s Invitation to Trust and

Mistrust Language," and "Line. On the Line. Lining Up. Lined With. Between the Lines. Bottom Line."

13. Fraser's commitment to feminist poetics has been well documented. In 1983, she founded the ground-breaking feminist journal *HOW(ever),* which for nine years published the work of experimental women poets and feminist scholars. See especially Linda Taylor, "'A Seizure of Voice': Language Innovation and a Feminist Poetics in the Works of Kathleen Fraser," and Cynthia Hogue, "'I am not of that feather': Kathleen Fraser's Postmodernist Poetics."

14. Poets whom Fraser names as working in this tradition include Barbara Guest, Susan Howe, Hannah Weiner, Myung Mi Kim, Dale Going, and Laura Moriarty.

15. For a discussion of Fraser's early work and influences, see Keller, "'Just one of / the girls:— / normal in the extreme': Experimentalists-To-Be Starting Out in the 1960s."

16. Even when her ekphrastic work has an explicitly feminist theme or subject, this content is deployed through formal exploration and aleatoric methods. For example, in "La La at the Cirque Fernando, Paris," which addresses Degas's portrait of the demimondaine, Fraser devises a "matrix" of words derived from a typo to reveal, as her note explains, "a core lexicon or set of word clues needed by La La in order for her to come into possession of her own voice and her autonomy from 'the boss'" (*il cuore* 196).

17. Fraser shares Levine's interest in copying: Levine writes, "I like a situation where notation becomes content and style. All the different manifestations equally represent the work" (qtd. in McShine 140). See, for example, Levine's 1994 series of photographs *After Van Gogh,* in which she makes pictures that "are really ghosts of ghosts; their relationship to the original images is tertiary, i.e. three or four times removed. By the time a picture becomes a bookplate it's already been rephotographed several times" (140).

18. Fraser's explorations intersect with Julia Kristeva's discussion, in "Giotto's Joy," of the painter's "experiments in architecture and color (his translation of instinctual drives into colored surface)" (210). Although Fraser does not name Kristeva as an explicit source for this particular poem, her approach has affinities with Kristeva's emphasis on "word-presentations" (a concept she borrows from Freud) as a category between the perceptual and the verbal: "Word-presentations would then be doubly linked to the body. First, as representations of an 'exterior' object denoted by the word, as well as representations of the [instinctual drive's] pressure itself, which, although intraorganic, nevertheless relates the speaking subject to the object. Second, as representations of an 'interior object,' an internal perception, an eroticization of the body proper during the act of formulating the word as a symbolic element" (217).

19. The lines in Italian are "sappi che 'l mio vicin Vitalïano / sederà *qui* dal mio sinistro fianco" (*Inferno* 17.68–9, my emphasis).

20. Most translations of these lines arrange the noun and adjective elements with only slight variation. Mandelbaum renders the image as "a goose more white than butter" (153). Stanley Plumly, in an edition of poets' translations of Dante's *Inferno* compiled by Daniel Halpern (1993), translates it "a white goose whiter than whey" (77). Robert Pinsky's version describes the goose "in a color whiter than butter" (173).

21. Carson is a professor of classics at McGill University in Montreal.

22. In an interview, Carson assented to the term "collage" as descriptive of her poetics, and likened her process to "Painting with thoughts and facts" (Interview 13). The visual arts also supply her with subject matter and themes: "Hopper: *Confessions*" and "Giotto Shot List" from *Men in the Off Hours* (2000) are ekphrastic studies of particular works, and the first "Tango" of *The Beauty of the Husband* (2001) addresses Duchamp's *The Bride Stripped Bare by Her Bachelors*.

23. Carson often juxtaposes historical or mythical material with the present day. In *Autobiography of Red* (1998), a contemporary teenager who is also the winged monster Geryon has an affair with Herakles. In "Mimnermos: The Brainsex Paintings" (1995), translations of Greek lyric poetry are combined with contemporary details (Rae 25). Many others of her poems make use of classical scholarship and allusions in contemporary contexts. For a discussion of these juxtapositions as hinging on matters of gender and genre, see Rae. For a less sympathetic account of Carson's "offbeat hiccups of anachronistic intrusion," see D. Ward (14).

24. The "prose-like" qualities of Carson's poems have provoked favorable and unfavorable critical responses. I have opted not to rehearse the usual critical fretting about whether her poems are "chopped up prose" or "playing tennis with the net down," and instead focused on the local effects of the lines and stanzas of this poem. For an excellent discussion of this issue, and an instructive comparison of two versions of one of Carson's poems, one lineated and one in prose, see John D'Agata's review of *Men in the Off Hours*. For a useful overview of Carson's "hybrids of poetic and critical forms," see Wahl (181).

25. In an interview, Carson remarked on the complexities of the notion of an "autobiographical" subject for poetry, commenting "That's a big space, the 'I'" (Interview 18). The "I" in her poems is frequently capacious, often multiple, seldom easily defined.

26. My reading of this poem focuses on its doublings, but it is also fruitful to consider its triangular patterns, patterns that recur in Carson's work as figures for desire, as Jennings has described (922). For example, the poem presents not only two temporal "centers of vision"—Perugino and the modern-day painter—but a triangle of desire that includes the poet-speaker (or Carson herself), the painter, and Anna—the object whom they both call into being, seek, and desire. Carson identifies the archetype of this triangu-

lar figure for eros in Sappho: "three points of transformation on a circuit of possible relationship, electrified by desire so that they touch not touching" (*Eros the Bittersweet* 16). In "Canicula di Anna," the speaker desires Anna and simultaneously transforms herself into Anna in a manner that looks forward to "Irony Is Not Enough: Essay on My Life as Catherine Deneuve (2nd draft)" (*Men in the Off Hours* [2000]). In this later poem, Deneuve is a classics professor who becomes infatuated with a female student. In "Canicula di Anna," the suggestion of desire between women is less apparent, deflected because the speaker morphs into "Perugino." Carson's triangulations recall René Girard's theory of "triangular desire" in the novel, although she does not, to my knowledge, refer to Girard directly in her writings. She shares his emphasis on "desire according to the Other" (5), and on "the double role of model and obstacle played by the mediator" (42).

27. Heidegger makes this etymological link to be able to claim that the ancients' understanding of Being was temporal (Being as Presence in the Present). I thank Troy Thibodeaux for his assistance on this point.

28. See histories of Perugia by Johnstone and Banker.

Bibliography

Adams, Pat, ed. *With a Poet's Eye: A Tate Gallery Anthology.* London: Tate Gallery, 1986.

Adorno, Theodor W. "Valéry Proust Museum." *Prisms.* Trans. Samuel and Shierry Weber. Cambridge, MA: The MIT P, 1981. 173–86.

Aisenberg, Katy. *Ravishing Images: Ekphrasis in the Poetry and Prose of William Wordsworth, W. H. Auden and Philip Larkin.* New York: Peter Lang, 1995.

Albers, Josef. *Interaction of Color.* New Haven: Yale UP, 1975.

Allen, Donald, ed. *The Collected Poems of Frank O'Hara.* New York: Knopf, 1971.

———, ed. [1960] *The New American Poetry 1945–1960.* Berkeley: U of California P, 1999.

Althusser, Louis. "Ideology and Ideological State Apparatuses." In *Lenin and Philosophy and Other Essays.* Trans. Ben Brewster. London: NLB, 1971. 123–73.

Altieri, Charles. "John Ashbery and the Challenge of Postmodernism in the Visual Arts." *Critical Inquiry* 14 (1988): 805–830.

———. *Self and Sensibility in Contemporary American Poetry.* Cambridge: Cambridge UP, 1984.

Altshuler, Bruce. *The Avant-Garde in Exhibition: New Art in the 20th Century.* New York: Harry N. Abrams, 1994.

Ashbery, John. *And the Stars Were Shining.* New York: Farrar, Straus and Giroux, 1994.

———. *April Galleons.* New York: Viking, 1987.

———. *As We Know.* New York: Viking, 1979.

———. *Can You Hear, Bird.* New York: Farrar, Straus and Giroux, 1995.

———. *Chinese Whispers.* New York: Farrar, Straus and Giroux, 2002.

———. *Flow Chart.* New York: Knopf, 1991.

———. Foreword. *Self-Portrait in a Convex Mirror.* San Francisco: Arion, 1984.

———. *Girls on the Run.* New York: Farrar, Straus and Giroux, 1999.

———. *Hotel Lautréamont.* New York: Knopf, 1992.

———. *Houseboat Days.* [1977] New York: Farrar, Straus and Giroux, 1999.

——. *The Mooring of Starting Out: The First Five Books of Poetry.* Hopewell, NJ: Ecco, 1997.

——. *Reported Sightings: Art Chronicles 1957–1987.* Ed. David Bergman. Cambridge, MA: Harvard UP, 1991.

——. *Selected Poems.* New York: Penguin, 1985.

——. *Self-Portrait in a Convex Mirror.* New York: Penguin, 1976.

——. *Wakefulness.* New York: Farrar, Straus and Giroux, 1998.

——. *A Wave.* New York: Viking, 1984.

——. *Where Shall I Wander.* New York: Ecco, 2005.

——. *Your Name Here.* New York: Farrar, Straus and Giroux, 2000.

Ashbery, John, and Kenneth Koch. "Death Paints a Picture." *ARTnews* 57.5 (Sept. 1958): 24, 63.

Bal, Mieke. *Looking In: The Art of Viewing.* Amsterdam: G+B Arts International, 2001.

Balken, Debra Bricker, ed. *Philip Guston's Poem-Pictures.* Seattle: U of Washington P, 1994.

Bang, Mary Jo. *The Eye Like a Strange Balloon.* New York: Grove, 2004.

Banker, James R. "The Social History of Perugia in the Time of Perugino." In *Pietro Perugino: Master of the Italian Renaissance.* Ed. Joseph Becherer. Grand Rapids, MI: The Grand Rapids Art Museum, 1997. 37–51.

Barthes, Roland. *Image, Music, Text.* New York: Hill and Wang, 1977.

Beach, Christopher. *Poetic Culture: Contemporary Poetry Between Community and Institution.* Evanston, IL: Northwestern UP, 1999.

Benjamin, Walter. *Illuminations.* New York: Schocken, 1969.

Bennett, Tony. *The Birth of the Museum: History, Theory, Politics.* New York: Routledge, 1995.

Benveniste, Émile. *Problems in General Linguistics.* Trans. Mary Elizabeth Meek. Coral Gables: U of Miami P, 1971.

Bergman, David. "Choosing Our Fathers: Gender and Identity in Whitman, Ashbery and Richard Howard." *American Literary History* 1 (1989): 383–403.

Bernstein, Charles. *A Poetics.* Cambridge, MA: Harvard UP, 1992.

Bishop, Elizabeth. *The Complete Poems 1927–1979.* New York: Farrar, Straus and Giroux, 1983.

——. *One Art: Letters.* Ed. Robert Giroux. New York: Farrar, Straus and Giroux, 1994.

Bourdieu, Pierre. *The Field of Cultural Production.* Ed. Randal Johnson. New York: Columbia UP, 1993.

Bourdieu, Pierre, and Alain Darbel. [1969] *The Love of Art: European Art Museums and Their Public.* Trans. Caroline Beattie and Nick Merriman. Stanford: Stanford UP, 1990.

Bourdieu, Pierre, and Hans Haacke. *Free Exchange.* Stanford: Stanford UP, 1995.

Brouwer, Joel. "Clio Rising." *Parnassus* 26.1 (2001): 291–307.

Bryson, Norman. Introduction. *Looking In: The Art of Viewing.* By Mieke Bal. Amsterdam: G+B Arts International, 2001. 1–39.

Bürger, Peter. *Theory of the Avant-Garde.* Minneapolis: U of Minnesota P, 1984.

Campo, Rafael. "Making Sense of the Currency on Line for Le Musée Picasso." *Ploughshares* 31.1 (Spring 2005): 21.

Carroll, Paul. *The Poem in Its Skin.* New York: Follett, 1968.

Carson, Anne. "A _____ with Anne Carson." Interview with John D'Agata. *Iowa Review* 27.2 (Summer/Fall 1997): 1–22.

——. *Autobiography of Red.* New York: Vintage, 1998.

——. *The Beauty of the Husband.* New York: Knopf, 2001.

——. *Eros the Bittersweet.* [1986] Normal, IL: Dalkey Archive, 1998.

——. *Glass, Irony, and God.* New York: New Directions, 1995.

——. *Men in the Off Hours.* New York: Knopf, 2000.

——. *Plainwater.* [1995] New York: Vintage, 2000.

Chiasson, Dan. "Him Again: John Ashbery." *Raritan* 21 (2001): 139–145.

Chinitz, David. "'Arm the Paper Arm': Kenneth Koch's Postmodern Comedy." In Diggory and Miller, 311–326.

Clampitt, Amy. *The Collected Poems of Amy Clampitt.* New York: Knopf, 1999.

Clark, T. J. *Farewell to an Idea: Episodes from a History of Modernism.* New Haven: Yale UP, 1999.

Clifford, James. *The Predicament of Culture.* Cambridge, MA: Harvard UP, 1988.

Conoley, Gillian. Rev. of *Try,* by Cole Swensen. *Boston Review* 24.5 (Oct./Nov. 1999): http://bostonreview.net/BR24.5/conoley.html.

Costello, Bonnie. "Charles Wright's *Via Negativa:* Language, Landscape, and the Idea of God." *Contemporary Literature* 42.2 (2001): 325–46.

Crimp, Douglas. *On the Museum's Ruins.* With photographs by Louise Lawler. Cambridge, MA: The MIT P, 1993.

D'Agata, John. Rev. of *Men in the Off Hours,* by Anne Carson. *Boston Review* 25.3 (Summer 2000): http://bostonreview.net/BR25.3/dagata.html.

Dante Alighieri. *The Divine Comedy: Inferno.* Trans. Allen Mandelbaum. New York: Bantam, 1982.

——. *The Inferno of Dante.* Trans. Robert Pinsky. New York: Farrar, Straus and Giroux, 1994.

Davidson, Michael. "Ekphrasis and the Postmodern Painter Poem." *The Journal of Aesthetics and Art Criticism* 42.1 (1983): 69–79.

da Vinci, Leonardo. *The Literary Works of Leonardo da Vinci.* Ed. Jean Paul Richter. New York: Oxford UP, 1939.

Diggory, Terence and Stephen Paul Miller, eds. *The Scene of My Selves: New Work on the New York School Poets.* Orono, ME: The National Poetry Foundation, 2001.

Dixon, A. Commentary on Joan Mitchell's *White Territory.* University of Michigan Museum of Art, Twentieth-Century Gallery Installation, June 1999.

Donaldson, Jeffery. "Going Down in History: Richard Howard's *Untitled Subjects* and James Merrill's *The Changing Light at Sandover.*" *Salmagundi* 76–77 (1987–88): 175–202.

Donnelly, Timothy. *Twenty-seven Props for a Production of Eine Lebenszeit.* New York: Grove, 2003.

Donoghue, Denis. *The Practice of Reading.* New Haven: Yale UP, 1998.

Duncan, Carol. *Civilizing Rituals: Inside Public Art Museums.* New York: Routledge, 1995.

Dunn, Allen. "Who Needs a Sociology of the Aesthetic? Freedom and Value in Pierre Bourdieu's *Rules of Art.*" *boundary 2* 25.1 (1998): 87–110.

DuPlessis, Rachel Blau. "'Corpses of Poesy': Some Modern Poets and Some Gender Ideologies of Lyric." In Keller and Miller, 69–95.

Edelman, Lee. "The Pose of Imposture: Ashbery's 'Self-Portrait in a Convex Mirror.'" *Twentieth Century Literature* 32.1 (1986): 95–114.

Epstein, Andrew. "'I want to be at least as alive as the vulgar': Frank O'Hara's Poetry and the Cinema." In Diggory and Miller, 93–121.

Ferguson, Suzanne. "Crossing the Delaware with Larry Rivers and Frank O'Hara: the post-modern hero at the Battle of Signifiers." *Word & Image* 2.1 (1986): 27–32.

Fineberg, Jonathan. *Art since 1940: Strategies of Being.* Englewood Cliffs, NJ: Prentice Hall, 1995.

Fitzpatrick, Kathleen. "The Exhaustion of Literature: Novels, Computers, and the Threat of Obsolescence." *Contemporary Literature* 43.3 (2002): 518–59.

Fraser, Kathleen. "The Blank Page: H.D.'s Invitation to Trust and Mistrust Language." In *H.D. and Poets After.* Ed. Donna Krolik Hollenberg. Iowa City, IA: U of Iowa P, 2000. 163–171.

———. *il cuore : the heart: Selected Poems 1970–1995.* Hanover, NH: Wesleyan UP, 1997.

———. "An Interview with Kathleen Fraser." With Cynthia Hogue. *Contemporary Literature* 39.1 (1998): 1–26.

———. "Line. On the Line. Lining Up. Lined With. Between the Lines. Bottom Line." In *The Line in Postmodern Poetry.* Ed. Robert Frank and Henry Sayre. Chicago: U of Illinois P, 1988. 152–74.

———. "Translating the Unspeakable: Visual Poetics, as Projected through Olson's 'Field' into Current Female Writing Practice." In *Moving Borders: Three Decades of Innovative Writing by Women.* Ed. Mary Margaret Sloan. Jersey City, NJ: Talisman, 1998. 642–54.

———. *when new time folds up.* Minneapolis: Chax Press, 1993.

Fulton, Alice. *Felt.* New York: Norton, 2001.

———. *Palladium.* Urbana: U of Illinois P, 1986.

———. *Sensual Math.* New York: Norton, 1995.

Genette, Gérard. *Seuils.* Paris: Seuil, 1987.

Gibson, Stephen. "The Battle of Lepanto." *Ploughshares* 31.1 (Spring 2005): 37–8.

Girard, René. *Deceit, Desire, and the Novel: Self and Other in Literary Structure.* Trans. Yvonne Freccero. Baltimore: The Johns Hopkins UP, 1966.

Golding, Alan. *From Outlaw to Classic: Canons in American Poetry.* Madison: U of Wisconsin P, 1995.

Graham, Jorie. *The Dream of the Unified Field.* Hopewell, NJ: Ecco, 1995.

Greger, Debora. *Western Art.* New York: Penguin, 2004.

Grunfeld, Frederic V. *Rodin: A Biography.* New York: Da Capo P, 1987.

Guest, Barbara. Interview with Mark Hillringhouse. *American Poetry Review* 21.4 (1992): 23–30.

——. *Selected Poems.* Los Angeles: Sun and Moon P, 1995.

Guillory, John. *Cultural Capital: The Problem of Literary Canon Formation.* Chicago: U of Chicago P, 1993.

Hagstrum, Jean H. *The Sister Arts: The Tradition of Literary Pictorialism and English Poetry from Dryden to Gray.* Chicago: U of Chicago P, 1958.

Hall, Donald, Robert Pack, and Louis Simpson, eds. *New Poets of England and America.* New York: Meridian, 1957.

Halpern, Daniel, ed. *Dante's Inferno: Translations by 20 Contemporary Poets.* Hopewell, NJ: Ecco, 1993.

Harper, Phillip Brian. *Private Affairs: Critical Ventures in the Culture of Social Relations.* New York: New York UP, 1999.

Harvey, Matthea. *Pity the Bathtub Its Forced Embrace of the Human Form.* Farmington, ME: Alice James, 2000.

Hayden, Robert. *Collected Poems.* Ed. Frederick Glaysher. New York: Liveright, 1985.

——. *Words in the Mourning Time.* New York: October House, 1970.

Hecht, Anthony. "At the Frick." In *The Gazer's Spirit.* Ed. John Hollander. Chicago: U of Chicago P, 1995. 259.

——. *On the Laws of the Poetic Art.* Princeton: Princeton UP, 1995.

Heffernan, James A. W. *Museum of Words: The Poetics of Ekphrasis from Homer to Ashbery.* Chicago: U of Chicago P, 1994.

Heidegger, Martin. *Being and Time.* Trans. John MacQuarrie and Edward Robinson. San Francisco: Harper, 1962.

Herd, David. *John Ashbery and American Poetry.* New York: Palgrave, 2000.

Hirsch, Edward. "From 'The Visionary Poetics of Philip Levine and Charles Wright.'" In *The Columbia History of American Poetry.* Ed. Jay Parini. New York: Columbia UP, 1994. 777–805.

——, ed. *Transforming Vision: Writers on Art.* Boston: Bulfinch for the Art Institute of Chicago, 1994.

Hogue, Cynthia. "'I am not of that feather': Kathleen Fraser's Postmodernist Poetics." In *H.D. and Poets After.* Ed. Donna Krolik Hollenberg. Iowa City, IA: U of Iowa P, 2000. 172–83.

Hollander, John. *The Gazer's Spirit: Poems Speaking to Silent Works of Art.* Chicago: U of Chicago P, 1995.

Hollander, John, and Joanna Weber, eds. *Words for Images: A Gallery of Poems.* New Haven: Yale University Art Gallery, 2001.

Hooper-Greenhill, Eilean. *Museums and the Shaping of Knowledge.* New York: Routledge, 1992.

Hoover, Paul. "Fables of Representation: Poetry of the New York School." *American Poetry Review* (July/August 2002): 20–30.

——, ed. *Postmodern American Poetry: A Norton Anthology.* New York: Norton, 1994.

——. "The Postmodern Era: A Final Exam." *Chicago Review* 45.3–4 (1999): 108–11.

Horace. *The Art of Poetry.* Verse translation by Burton Raffel. Albany: State U of New York P, 1974.

Howard, Richard. *Alone with America: Essays on the Art of Poetry in the United States since 1950.* New York: Atheneum, 1969.

——. *The Damages.* Middletown, CT: Wesleyan UP, 1967.

——. *Fellow Feelings.* New York: Atheneum, 1976.

——. "In Loco Parentis, 1963." *The Georgia Review* 57.4 (Winter 2003): 734–6.

——. *Inner Voices: Selected Poems 1963–2003.* New York: Farrar, Straus and Giroux, 2004.

——. Interview with Mary Jo Bang. *Boulevard* 15.3 (2000): 37–47.

——. *Like Most Revelations.* New York: Pantheon Books, 1994.

——. *Lining Up.* New York: Atheneum, 1984.

——. *Misgivings.* New York: Atheneum, 1979.

——. *No Traveller.* New York: Knopf, 1989.

——. *Paper Trail: Selected Prose 1965–2003.* New York: Farrar, Straus and Giroux, 2004.

——. "Susan Wheeler." *Boston Review* 28.5 (2003): 54.

——. *Talking Cures.* New York: Turtle Point P, 2002.

——. *Trappings.* New York: Turtle Point P, 1999.

——. *Untitled Subjects.* New York: Atheneum, 1969.

Huyssen, Andreas. *Twilight Memories: Marking Time in a Culture of Amnesia.* New York: Routledge, 1995.

Jennings, Chris. "The Erotic Poetics of Anne Carson." *University of Toronto Quarterly* 70.4 (Fall 2001): 923–36.

Johnson, Randal. Introduction. *The Field of Cultural Production.* By Pierre Bourdieu. New York: Columbia UP, 1993. 1–25.

Johnstone, Mary. *Perugia and her People.* Perugia: Grafica, 1956.

Keller, Lynn. "'Just one of / the girls:— / normal in the extreme'": Experimentalists-To-Be Starting Out in the 1960s." *Differences: A Journal of Feminist Cultural Studies* 12.2 (2001): 47–69.

——. "The '*Then Some* Inbetween': Alice Fulton's Feminist Experimentalism." *American Literature* 71.2 (1999): 311–340.

Keller, Lynn, and Cristanne Miller, eds. *Feminist Measures: Soundings in Poetry and Theory.* Ann Arbor: U of Michigan P, 1994.

Kellogg, David. "The Self in the Poetic Field." *Fence* 3.2 (2000–2001): 97–108.

Kelly, Lionel, ed. *Poetry and the Sense of Panic: Critical Essays on Elizabeth Bishop and John Ashbery.* Amsterdam: Rodopi, 2000.

Kincaid, James R. "Resist Me, You Sweet Resistible You." *PMLA* 118.5 (2003): 1325–33.

Kirchwey, Karl. *At the Palace of Jove.* New York: Putnam, 2002.

Koch, Kenneth. *The Art of the Possible: Comics Mainly without Pictures.* New York: Soft Skull P, 2004.

——. *The Gold Standard: A Book of Plays.* New York: Knopf, 1996.

——. *Hotel Lambosa.* Minneapolis: Coffee House Press, 1995.

——. Interview with Jordan Davis. *American Poetry Review* 25.6 (1996): 45–53.

——. *On the Great Atlantic Rainway: Selected Poems 1950–1988.* New York: Knopf, 1994.

Koethe, John. "The Absence of a Noble Presence." In Schultz, 83–90.

Krieger, Murray. *Ekphrasis: The Illusion of the Natural Sign.* Baltimore: Johns Hopkins UP, 1992.

Kristeva, Julia. *Desire in Language.* New York: Columbia UP, 1980.

Kunitz, Stanley. *Passing Through: The Later Poems New and Selected.* New York: Norton, 1995.

Kwon, Miwon. *One Place After Another: Site-Specific Art and Locational Identity.* Cambridge, MA: The MIT P, 2002.

Leckie, Ross. "Art, Mimesis, and John Ashbery's 'Self-Portrait in a Convex Mirror.'" *Essays in Literature* 19.1 (1992): 114–31.

Lehman, David, ed. *Beyond Amazement: New Essays on John Ashbery.* Ithaca: Cornell UP, 1980.

——. *The Last Avant-Garde: The Making of the New York School of Poets.* New York: Doubleday, 1998.

Lessing, Gotthold Ephraim. [1766] *Laocoön: An Essay on the Limits of Painting and Poetry.* Trans. Edward Allen McCormick. Baltimore: Johns Hopkins UP, 1984.

Lisk, Thomas. "An Ashbery Primer." In Diggory and Miller, 35–49.

Logan, William. "Late Callings." *Parnassus* 18–19 (1993): 317–27.

Loizeaux, Elizabeth Bergmann. "Ekphrasis and textual consciousness." *Word & Image* 15.1 (1999): 76–96.

Lolordo, Nick. "Charting the Flow: Positioning John Ashbery." *Contemporary Literature* 42 (2001): 750–774.

Longenbach, James. "Ashbery and the Individual Talent." *American Literary History* 9.1 (1997): 103–27.

——. "Richard Howard's Modern World." *Salmagundi* 108 (1995): 140–63.

Lundquist, Sara. "Another Poet Among Painters: Barbara Guest with Grace Hartigan and Mary Abbott." In Diggory and Miller, 245–64.

———. "Reverence and Resistance: Barbara Guest, Ekphrasis, and the Female Gaze." *Contemporary Literature* 38.2 (1997): 260–86.

Malraux, André. *Museum without Walls.* Trans. Stuart Gilbert and Francis Price. New York: Doubleday, 1967.

Mazzaro, Jerome. "Fact and Matter: Richard Howard's 'Even in Paris.'" *Salmagundi* 76–77 (1987–88): 159–74.

McCorkle, James. "The Demands of Reading: Mapping, Travel and Ekphrasis in the Poetry from the 1950s of John Ashbery and Elizabeth Bishop." In Diggory and Miller, 67–89.

McHale, Brian. "How (Not) to Read Postmodernist Long Poems: The Case of Ashbery's 'The Skaters.'" *Poetics Today* 21.3 (Fall 2000): 561–590.

McShine, Kynaston. *The Museum as Muse: Artists Reflect.* New York: The Museum of Modern Art, 1999.

Meek, Sandra Lea. "The Politics of Poetics: Creative Writing Programs and the Double Canon of Contemporary Poetry." In *Canon vs. Culture: Reflections on the Current Debate.* Ed. Jan Gorak. New York: Garland, 2001. 81–102.

Miller, Cristanne. "'The Erogenous Cusp,' or Intersections of Science and Gender in Alice Fulton's Poetry." In Keller and Miller, 317–43.

Miller, Stephen Paul. "'Self-Portrait in a Convex Mirror,' the Watergate Affair, and Johns's Crosshatch Paintings: Surveillance and Reality-Testing in the Mid-Seventies." *boundary 2* 20.2 (1993): 84–115.

Mitchell, W. J. T. "Ekphrasis and the Other." In *Picture Theory: Essays on Verbal and Visual Representation.* Chicago: U of Chicago P, 1994. 151–181.

———. *Iconology: Image, Text, Ideology.* Chicago: U of Chicago P, 1986.

Moramarco, Fred. "Coming Full Circle: John Ashbery's Later Poetry." In Schultz, 38–59.

Mueller, Robert. "John Ashbery and the Poetry of Consciousness: 'Self-Portrait in a Convex Mirror.'" *The Centennial Review* 40.3 (1996): 561–72.

O'Doherty, Brian. *Inside the White Cube: The Ideology of the Gallery Space.* Berkeley: U of California P, 1999.

Oehlschlaeger, Fritz. "Robert Hayden's Meditation on Art: The Final Sequence of *Words in the Mourning Time.*" *Black American Literature Forum* 19.3 (1985): 115–119.

O'Hara, Frank. *Art Chronicles, 1954–1966.* New York: George Braziller, 1975.

———. *The Collected Poems of Frank O'Hara.* Ed. Donald Allen. New York: Knopf, 1971.

Paul, Catherine. *Poetry in the Museums of Modernism: Yeats, Pound, Moore, Stein.* Ann Arbor: U of Michigan P, 2002.

Pelton, Theodore. "Kenneth Koch's Poetics of Pleasure." In Diggory and Miller, 327–44.

Perloff, Marjorie. *21st-Century Modernism: The "New" Poetics.* Malden, MA: Blackwell, 2002.

———. "Can(n)on to the Right of Us, Can(n)on to the Left of Us: A Plea for Difference." *New Literary History* 18.3 (1987): 633–56.

———. *Frank O'Hara: Poet Among Painters.* Austin: U of Texas P, 1979.

———. "Normalizing John Ashbery." *Jacket* 2 (1997): http://jacketmagazine.com/02/perloff02.html.

Rae, Ian. "'Dazzling Hybrids': The Poetry of Anne Carson." *Canadian Literature* 166 (Autumn 2000): 17–41.

Ramke, Bin. "Reading Off the Wall: Recent Books by Richard Howard." *Denver Quarterly* 30.2 (1995): 125–9.

Rankine, Claudia. *Don't Let Me Be Lonely: An American Lyric.* Saint Paul, MN: Graywolf P, 2004.

Rasula, Jed. *The American Poetry Wax Museum: Reality Effects, 1940–1990.* Urbana, IL: National Council of Teachers of English, 1996.

Rich, Adrienne. *Collected Early Poems 1950–1970.* New York: Norton, 1993.

Rifkin, Libbie. "Making It / New: Institutionalizing Postwar Avant-Gardes." *Poetics Today* 21.1 (2000): 129–50.

Rilke, Rainer Maria. *Auguste Rodin.* New York: Haskell, 1974.

Ross, Andrew. "Taking the Tennis Court Oath." In Schultz, 193–210.

Ruskin, John. [1854] *Giotto and His Works in Padua.* Boston: Longwood P, 1977.

Russell, John. *Max Ernst.* London: Thames and Hudson, 1967.

Santos, Sherod. "Girl Falling Asleep in the Museum Gardens." *The Yale Review* 93.2 (April 2005): 24–5.

Schjeldahl, Peter. "Folks." *The New Yorker.* January 14, 2002: 88–9.

Schneemann, Carolee. *More than Meat Joy: Complete Performance Works & Selected Writings.* Ed. Bruce McPherson. New Paltz: Documentext, 1979.

Schultz, Susan M., ed. *The Tribe of John: Ashbery and Contemporary Poetry.* Tuscaloosa: U of Alabama P, 1995.

Scott, Grant. "Copied with a difference: *ekphrasis* in William Carlos Williams' *Pictures from Brueghel.*" *Word & Image* 15.1 (1999): 63–75.

Shaughnessy, Brenda. *Interior with Sudden Joy.* New York: Farrar, Straus and Giroux, 1999.

Sherman, Daniel J. "Quatremère/Benjamin/Marx: Art Museums, Aura, and Commodity Fetishism." In *Museum Culture: Histories, Discourses, Spectacles.* Ed. Daniel J. Sherman and Irit Rogoff. Minneapolis: U of Minnesota P, 1994. 123–43.

Sleigh, Tom. "The Mouth." *TriQuarterly* 120 (2004): 149–50.

Spies, Werner. *Max Ernst 1950–1970.* New York: Harry N. Abrams, 1971.

Spurr, David. "Kenneth Koch's 'Serious Moment.'" In Diggory and Miller, 345–56.

Stamelman, Richard. "Critical Reflections: Poetry and Art Criticism in Ashbery's 'Self-Portrait in a Convex Mirror.'" *New Literary History* 15.3 (1984): 607–30.

Staniszewski, Mary Anne. *The Power of Display: A History of Exhibition Installations at the Museum of Modern Art.* Cambridge, MA: The MIT P, 1998.

Stevens, Wallace. *The Collected Poems.* New York: Vintage Books, Random House, Inc., 1990.

———. *Letters of Wallace Stevens.* Ed. Holly Stevens. Berkeley: U of California P, 1996.

Suggs, M. Jack, Katharine Doob Sakenfeld, and James R. Mueller, eds. *The Oxford Study Bible.* New York: Oxford UP, 1992.

Sweet, David. "Parodic Nostalgia for Aesthetic Machismo: Frank O'Hara and Jackson Pollock." *Journal of Modern Literature* 23.3–4 (2000): 375–91.

———. "'And *Ut Pictura Poesis* Is Her Name': John Ashbery, the Plastic Arts, and the Avant-Garde." *Comparative Literature* 50.4 (1998): 316–32.

Swensen, Cole. *New Math.* New York: William Morrow, 1988.

———. *Such Rich Hour.* Iowa City: U of Iowa P, 2001.

———. "To Writewithize." *American Letters & Commentary* 13 (2001): 122–27.

———. *Try.* Iowa City: U of Iowa P, 1999.

Sylvester, David. *Magritte: The Silence of the World.* New York: Harry N. Adams, 1992.

Taylor, Linda. "'A Seizure of Voice': Language Innovation and a Feminist Poetics in the Works of Kathleen Fraser." *Contemporary Literature* 33.2 (1992): 337–72.

Thévoz, Michel. "The Strange Hell of Beauty. . . ." In *Darger: The Henry Darger Collection at the American Folk Art Museum.* Ed. Brooke Davis Anderson. New York: Harry N. Abrams, 2001. 15–21.

Tillinghast, Richard, ed. *A Visit to the Gallery.* Ann Arbor: University of Michigan Museum of Art, 1997.

Valéry, Paul. [1923] "The Problem of Museums." In *Degas, Manet, Morisot.* Trans. David Paul. *The Collected Works in English.* Vol. 12. New York: Pantheon/Bollingen, 1960. 202–6.

Vasari, Giorgio. *Lives of the Artists.* Trans. George Bull. Vol. 1. New York: Penguin, 1987.

Vendler, Helen. *The Music of What Happens.* Cambridge, MA: Harvard UP, 1988.

———. "Reading and Hearing John Ashbery's 'Self-Portrait in a Convex Mirror.'" *Self-Portrait in a Convex Mirror.* San Francisco: Arion, 1984. N. pag.

Von Hallberg, Robert. *American Poetry and Culture 1945–1980.* Cambridge: Harvard UP, 1985.

Wahl, Sharon. "Erotic Sufferings: *Autobiography of Red* and Other Anthropologies." *Iowa Review* 29.1 (Spring 1999): 180–88.

Ward, David C. "Anne Carson: Addressing the Wound." *PN Review* 27.5/139 (May/June 2001): 13–16.

Ward, Geoff. "'Why, it's right there in the *procès verbal*': The New York School of Poets." *The Cambridge Quarterly* 21.3 (1992): 273–82.

Wheeler, Susan. *Ledger.* Iowa City: U of Iowa P, 2005.

Whitfield, Sarah. *Magritte.* London: The South Bank Centre, 1992.

Whitman, Walt. [1855] *Leaves of Grass.* New York: Penguin, 1959.

Wieners, John. *Selected Poems 1958–1984.* Ed. Raymond Foye. Santa Barbara, CA: Black Sparrow P, 1986.

Witek, Terri. *Fools and Crows.* Alexandria, VA: Orchises P, 2003.

Wolf, Leslie. "The Brushstroke's Integrity: The Poetry of John Ashbery and the Art of Painting." In Lehman, ed., 224–54.

Wright, Charles. *Negative Blue: Selected Later Poems.* New York: Farrar, Straus and Giroux, 2000.

Index